Network-Aware Security for Group Communications

T0137955

Yan Sun • Wade Trappe • K.J.R. Liu

Network-Aware Security for Group Communications

 Springer

Yan Sun

Department of Electrical
  & Computer Engineering
University of Rhode Island
4 East Alumni Avenue
Kingston, RI 02881
USA

Wade Trappe

WINLAB
Rutgers University
73 Brett Road
Piscataway, NJ 08854
USA

K.J.R. Liu

Department of Electrical
  & Computer Engineering
University of Maryland
8603 34th Avenue
College Park, MD 20740
USA

ISBN 978-1-4419-4335-4   e-ISBN 978-0-387-68848-0

Printed on acid-free paper.

9 8 7 6 5 4 3 2 1

springer.com

To Our Families

Preface

Over the past few decades, the role of computing has grown from being used mainly for scientific purposes, into being part of our everyday life, where it is used for purposes such as communication, entertainment, and device control in the state-of-the-art consumer products. The ubiquity of communication networks is facilitating the development of wireless and Internet applications aimed at allowing users to communicate and collaborate amongst themselves. Soon, group-oriented services will be customary-they will be essential for increasing productivity in the future workplace, and they will be integral to how we redefine our sense of community. Ultimately, these group-oriented services will be heterogeneous in nature, bringing together a diverse clientele of users with varying amounts of computing power and communication capabilities. However, before these group-oriented services can materialize, technologies must be developed to guarantee that the information and data exchanged in these group-scenarios are protected. In short, it is necessary to develop solutions that will make multi-user services trustworthy and secure.

Recently computing and networking research has shifted from the static model of the wired Internet towards the new and exciting "anytime-anywhere" service model of the mobile Internet. At the heart of the technologies facilitating such pervasive computing are recent advancements in wireless technologies that will provide the ubiquitous communication coverage that is so coveted by mobile services. Moreover, due to the fact that wireless devices can seamlessly blend into users' lives, it is easy to predict that future wireless networks will gradually become the primary interface for consumer applications. These group-oriented services will be popular as they will be essential for increasing productivity in the future workplace. Already the migration to mobile computing has started, and it appears that the market for mobile services, or "m-commerce", will succeed as recent estimates project m-commerce to grow to involve over 1 billion subscribers. In spite of the predicted success of the wireless market, there are several disruptive challenges lurking in the future that threaten the successful adoption of wireless services. Perhaps core amongst these challenges are two issues, namely, platform heterogeneity, and secure and trusted communications.

The first issue points to the fact that wireless systems appear to be shifting away from the single-platform model of the 1990's to a free-for-all mixture of technologies battling it out in unlicensed bands of spectrum. Even the broad umbrella of Beyond-3G and 4G systems, along with forward-planning 3G/WLAN interworking solutions, do not appear to be positioned to capture the broad heterogeneity that will be introduced when completely new classes of wireless systems, such as cognitive radios, mesh networks, and wireless personal area networks are deployed using newly-developed programmable radio technologies. Further, it can be expected that a diverse array of new media services will drive the mobile Internet, and new multimedia delivery devices, such as wireless audio-visual devices and the next evolution of wearable computing devices, will emerge as important new products complementing today's laptop computers and personal digital assistants, providing a revolutionary means to communicate and collaborate from anywhere at anytime.

The second hurdle facing wireless systems is security. Even for the existing wireless networks, security is often cited as a major technical barrier that must be overcome before widespread adoption of mobile services can occur. The increasingly popular "WiFi" or 802.11 wireless local-area network was initially based on a standard with relatively weak wireless security called WEP, resulting in major security concerns as the equipment was deployed in offices and homes. Further, emerging 3G cellular data services also have limited security capabilities. Moreover, it has become clear that end-to-end security solutions, which were originally designed for the wired Internet, have limited applicability to the unique problems associated with wireless networks. Add to this the foreseeable heterogeneity in devices and user profiles that emerging wireless networks will introduce, and it is evident that there is a need for research targeted at developing security solutions for next-generation mobile services.

One of the most suitable technologies for delivering data to groups of users is multicast networking. Multicasting has seen significant advancements recently, in both the underlying technology as well as the deployment of applications utilizing multicast technologies. Already there are multicast services that stream stock quotes, and provide video and audio on demand. The adaptation of multicast into commercial applications requires security functionalities, such as authentication, non-repudiation, and access control. Of these, access control is paramount as it is the first line of defense needed to protect the value of an application's data. A service provider may control access to content by encrypting the content using a key that is shared by all group members. The problem of access control becomes more difficult when the content is distributed to a group of users since membership will most likely be dynamic, with users joining and leaving the service for a variety of reasons, and therefore necessitating the ability to update keys.

Key management is accomplished either by using a centralized entity that is responsible for distributing keys to users, or by contributory protocols where legitimate members exchange information to agree upon a key. Typical group key management schemes seek to minimize either the amount of rounds needed in establishing the group key, or the size of the messages, and treat all users as identical. However, these approaches do not factor in the varying requirements of the users, the underlying network, or the application, and are therefore not well suited to provide solutions efficient for all users, for all networks, or for all types of applications. In particular, since many applications will involve a heterogeneous clientele consisting of group members with different computational capabilities, pricing plans, and bandwidth resources, these network-aware factors must be considered when designing an access control system.

The pervasiveness of computing has made it increasingly difficult to find any aspects of computing that are unaffected by issues from the underlying application and communication network. Applications must consider the requirements of the users and the underlying network conditions in order to provide a service that meets the demands of as many users as possible. A similar approach is needed for designing the security architecture for an application. In order to secure tomorrow's computing systems, it is essential to develop a network-aware framework that provides trustworthiness by jointly considering issues of computing and communications in dynamic, heterogeneous group environments.

Wireless multicasting will support many new multimedia applications, ranging from the broadcasting of media content for entertainment services, to video surveillance for remote monitoring applications, to multiparty "on-the-go" collaborations that will increase our productivity. Securing the next wave of wireless communications will require new strategies since traditional multicast security solutions are not targeted at addressing issues specific to emerging new applications such as wireless multimedia multicast services.

Before group-oriented wireless services can materialize, technologies must be developed to guarantee that the information and data exchanged are protected. In short, it is necessary to develop solutions that will make wireless multi-user multimedia services trustworthy and secure in the diverse wireless networks of the future. In order to accomplish this we have to have a better understanding with a holistic view of security solutions that address the following three topics:

- Access Control and Data Confidentiality serve as the first line of defense needed to protect the value of an application's data. A service provider may control access to content by encrypting the content using a key that is shared by all group members. The problem of access control for multicasts is challenging since group membership will most likely be dynamic, with users joining and leaving-necessitating the ability to update keys. However, traditional multicast key management schemes do not factor in the varying requirements of the users, the underlying network, or the application, and therefore are not adequately efficient for wireless multimedia multicast services.

- Service Authentication and Verification are important security issues for the media service. Traditional public key authentication is not suitable for wireless networks since many mobile devices will be low-powered, with limited computational and storage resources. Additionally, the strict delay requirements of multimedia data prevent popular delayed key disclosure techniques from being appropriate for wireless multimedia services. Together, these requirements necessitate the development of new classes of delay-sensitive authentication mechanisms for multimedia multicasting. An additional issue that is relevant for service validation is non-repudiation. Although non-repudiation is not typically studied in the context of multicast services, it is of particular importance for multimedia multicast services since the combination of advanced compression coding and best-effort wireless multicasting will not provide any guarantee of the quality of service delivered. It is important to both the service provider and the customers that mechanisms are available to irrefutably prove the quality of service delivered during a multimedia multicast service.

- Attack and Immunization Countermeasures are part of the security design cycle. The development of a suite of security protocols should involve an active phase of attacking the protocols in the suite as well as other protocols. The lessons learned by this effort give valuable insight into strengthening, or immunizing, the protocols to different types of attacks.

Throughout the discussion of these topics in this book, we take the viewpoint that the combination of content and wireless infrastructure introduces unique challenges that are not adequately addressed by generic multicast security solutions. This book presents the research results that have been undertaken by the authors in the past decade on security and reliability issues of group-based computing and communications. We hope our articulating point of the book– the network-aware approach toward security of group communications– can serve as an enlightening view for future development of wireless security.

Finally, we would like to acknowledge the assistance of the Army Research Office, whose University Research Initiatives has helped support the investigations behind many of the results that we present in this book. Additionally, we would like to express our thanks to the many people who have helped us in developing this book, including Yinian Mao, Min Wu, Yinian Mao, Jie Song, Wei Yu, and Qing Li.

Contents

1
Introduction

Communication technologies are rapidly maturing, and already the past decade has witnessed new forms of communication services being deployed. The deployment of various broadband communication technologies, such as digital subscriber line (DSL) and fiber optical communications, has led to a rapid price drop for bandwidth. Access networks, such as wireless local area networks (WLAN), are now commonplace and are rapidly evolving into metropolitan-style mesh networks.

Parallel to the rapid development of communication technologies, has been a surge in information applications. Multimedia content has become ubiquitous. Content editing software and hardware, such as digital cameras, are allowing for users to easily create. The availability of the Internet and the Web has encouraged artists, both professional and amateur, to share their creative expressions. This combination of application and communication technologies has created opportunities for new businesses to meet the growing global demand for information and entertainment.

Our communication infrastructures will continue to evolve. As bandwidth increases and costs for consumers to enter into the global Internet continue to decrease, geographical barriers will dissolve. As users are brought virtually closer to each other and made increasingly aware of each other, they will want to interact. Whether for good or for bad purposes, they will be able to share experiences that allow them to work or play together. Already, new commercial markets, such as interactive television and mobile video conferences, are on the horizon and promise to take advantage of the available bandwidth. It is no longer difficult to envision a future where users will personalize their experiences by interacting with each other. New forms of marketing and sales will open up, such as geocasting to consumers based on their location. Consumers with wireless appliances will receive messages advertising sales as they enter shopping centers, or walk down the street.

Although we traditionally think of communication as a point-to-point paradigm, the reality is that at many different levels communication is inherently group-oriented. In the example above, it is unreasonable to target a single individual with an advertisement, but instead it is far more effective to market towards a group of potential customers. Interactive games are not very interactive or entertaining without the social benefits of interacting with many other players.

Even should the primary communication be between two participants, there are many other communication operations that take place within a network that support this primary communication, and many of these operations are necessarily group-oriented. For example, in an ad hoc wireless network, where mobile devices communicate with each other via multihop routing, topology and route discovery procedures require broadcasting network-control packets across the network.

Group-oriented communications therefore represent both an important form of application traffic, as well as a critical form of control traffic. Such an important and valuable mode of communication not only needs to be reliable, but also secure. Although the effectiveness and reliability of group-oriented communication paradigms is important, paradigms like broadcast, multicast, gossip and flooding protocols are already well documented in a variety of references [1, 2].

Multicast communication is the most suitable method for delivering data to multiple entities due to its efficient usage of network resources. Over the Internet, for example, the recipients of a group communication are associated with a Class D IP address, and may receive messages sent to that address [2]. A server that desires to send communication to the group addresses messages with the group address and transmits a single copy of the message. It is the responsibility of the network and the multicast-enabled routers to deliver the message to the users. By sending only a single copy of the message on the network, the usage of server-side resources such as bandwidth and processing is reduced.

Instead of focusing on the communication protocols underlying group communication, this book focuses on issues related to the trustworthiness of group communications. The adaptation of multicast into commercial applications depends on the ability to control access to the communications. For example, consider a service provider that distributes streaming content, such as multimedia streams, to a group of paying users via a multicast technology. In such an application, the service provider must be able to ensure the availability of multicast data to privileged members while preventing unauthorized use of this data by non-privileged users. A service provider may control access to content by encrypting the content using a key that is shared by all group members. The problem of access control is made more difficult when the content is being distributed to a group of users since the membership will most likely be dynamic, with users joining and leaving the service for a variety of possible reasons. Upon changes in the membership, it is necessary to change the keys associated with the service.

Many applications will require that the communication amongst group members be protected from unwanted eavesdroppers. Corporate conferences, with members from different parts of the world, might contain industrial secrets that are in the best interests of the corporation to keep unknown to rivals. In order to protect the communication traffic, the information must be encrypted, requiring that the privileged parties share an encryption and decryption key.

The most appropriate framework for securing server-oriented content distribution is by using a centralized entity that is responsible for maintaining the integrity of the users' keys. The problem of maintaining access control is difficult when the content is being distributed to a group of users since the membership will most likely be dynamic with users joining and leaving the service. Unlike unicast communication, the departure of a group member does not imply the termination of the communication link. In addition, upon departing the service, users must be de-registered and prevented from obtaining future multicasts. Similarly, when new members join the service, it is desirable to prevent them from accessing past content.

The problem of key management for multicasts has seen recent attention in the literature, and several efficient schemes have been proposed with desirable communication properties. Most of these schemes, however, have not considered application-specific or network-specific properties that might affect the design of an access control system. As an example, multimedia data has rich properties, such as the capability to have information invisibly hidden in it and operate in a scalable or layered format, and these properties may be exploited to achieve an improved design of an access control system for multimedia multicasts.

The objective behind this book is to examine aspects of secure multi-user communications, and to highlight the aspects needed to tune secure group communications for a variety of future applications and networks.

1.1 Book Overview

This book focuses on secure group communication. At the most basic level, this deals with assuring that communications between more than two parties are confidential and that messages shared come from a source that is identifiable. However, a theme throughout this book is that such secure communication should be tailored to specific network scenarios or applications so that performance improvement can be achieved.

Looking forward, the book covers a range of topics in secure group communication, ranging from contributory to centralized key management, and from group communication for wireless networks to group communication for multimedia applications. A break down of the chapters is:

- Chapter 2 presents the fundamental issues associated with *centralized* key management schemes for group-oriented applications. In particular, we first overview the limitation in centralized multicast key distribution, then provide a survey of several existing approaches, and finally present a new framework for multicast key management. This new framework can handle dynamic group environments. This framework leads to efficient handling of dynamic member join, departure and transferal of access right.

- Chapter 3 presents *contributory* solutions to the key management problem for group-oriented applications. In this chapter, we first discuss the application scenarios in which contributory solutions are necessary or more favorable compared to centralized solutions. Then, the Diffie-Hellman (DH) protocol is reviewed followed by a discussion of several existing contributory key agreement schemes that employ or extend the DH protocol. Finally, a better approach, called butterfly scheme, is discussed in detail. The butterfly scheme can efficiently handle dynamic group members in a fully distributed manner. More importantly, it can accommodate heterogeneities in group members. When the group members have different capabilities and constrains, the scheme we describe is able to construct an adaptive key tree structure and can achieve nearly optimal performance.

- Chapter 4 examines the problem of optimizing rekeying cost in contributory key agreement protocols. Some performance lower bounds are derived through theoretical analysis. We also discuss two key agreement schemes that seek to achieve these lower bounds. Particularly, the first method, referred to as JET, uses a special join-tree/exit-tree topology and takes advantage of cost amortization. This method can significantly reduce the rekeying cost for user join. Inspired by JET, the second method consists of a new key tree structure, called PFMH, and a key agreement protocol, called PACK. This method only needs $O(1)$ rounds of two-party DH to handle user join and $O(\log n)$ rounds of two-party DH to handle user departure. This performance asymptotically approaches the performance lower bounds.

- Chapter 5 examines the key management problem in cellular wireless networks. As wireless connections become ubiquitous, consumers will desire to have multicast applications running on their mobile devices. In wireless environment where bandwidth is limited and transmission error rate is high, the design of key management schemes needs to put special attention on the transmission of the rekeying messages. In this chapter, we first discuss the targeting property of centralized key agreement schemes, and then present a topology-aware key management scheme. The topology-aware key management approach takes advantage of the targeting property as well as the broadcasting nature of the wireless media. By employing topology-awareness, the communication overhead associated with key updating can be significantly reduced. As a direct consequence, the reliability of key distribution can be greatly improved.

• Chapter 6 investigates the key management solutions to *multimedia multicast* applications. The demand for distributing multimedia to a large audience is a driving force for multicast research. Different from generic data, multimedia data has rich properties that can be used to facilitate the distribution of keying information. In this chapter, we examine the problem of managing keys for securing multimedia multicasts. Then, we show the importance of reducing the communication overhead associated with identifying which portion of a rekeying message is associated with each user. This communication overhead can be reduced by using a homogenized message format from which every user can perform a suitable operation to extract the new keying information. Finally, we identify and compare two ways to distribute keying information: media independent and media-dependent. In the media independent approach, keying information and multicast data are separated. In the media-dependent scheme, the keying information is embedded into the multimedia content, using data embedding techniques. The advantages of the media-dependent keying distribution is discussed and evaluated.

• Chapter 7 presents the hierarchical access control problem for group applications. Many group applications contain multiple related data streams and group members have different access privileges. These applications are prevalent in many scenarios. For example, multimedia applications often distribute data in a multi-layer coding format, and commercial multicast programs can contain several related services, such as weather, news, traffic and stock quote. In these scenarios, group members subscribe to different data steams, or possibly multiple of them. In other words, the access control mechanism needs to supports multi-level access privilege. This is referred to as *hierarchical group access control*. In this chapter, we formulate the hierarchical group access control problem and present efficient solutions in both centralized and contributory environments.

• Chapter 8 investigates a new aspect in designing group key management protocols. The design of current key management schemes focuses on maintaining key secrecy and reducing overhead associated with key updating. However, it is found that key management can disclose information about dynamic group membership to both insiders and outsiders. In other words, while the content of group communication is protected by encryption using the secret keys, group dynamic information is disclosed through key management. *Group dynamic information* (GDI) is the information that describes the dynamic group membership, including the number of users in a group as a function of time, and the number of joining or departing users in a time interval. In many secure group applications, group dynamic information should be kept confidential. In this chapter, we find possible methods for an unauthorized party to obtain GDI from key management schemes, and then investigate how to protect GDI from insiders and outsiders by making modifications to existing key management approaches.

- Chapter 9 addresses the multicast authentication problem, with an emphasis on reducing authenticator delay and improving resilience against denial of service (DoS) attacks. One security service that has been difficult to provide for multicast is authentication. In this chapter, we first review representative multicast source authentication schemes, and give a brief overview of the conventional TESLA scheme. Then, we present the Staggered TESLA scheme, which can reduce the authentication delay. We also derive theoretical guidelines for buffer requirements and discuss the tradeoffs involved in Staggered TESLA. The objective is to present strategies that reduce the delay associated with multicast authentication, make more efficient usage of receiver-side buffers, make delayed key disclosure authentication more resilient to buffer overflow denial of service attacks, and allow for multiple levels of trust in authentication.

- Chapter 10 represents an application of multicast authentication techniques to provide important authentication services for wireless networks where the participating devices have resource limitations. In particular, remote sensing applications are becoming an increasingly important area for research and development due to the critical need for applications that will perform environmental monitoring, provide security assurance, assist in healthcare services and facilitate factory automation. The authentication of the data source as well as the data is of critical concern since adversaries might attempt to capture sensors and tamper with sensor data. In this chapter, we first introduce some basic concepts in hierarchical sensor networks, and then describe TESLA certificates, which represent a modification of delayed-key disclosure multicast authentication to achieve a certificate framework. Finally, we describe a distributed light-weight framework for authentication that involves network nodes requesting trust references from neighboring nodes in order to establish the trust relationships needed for network authentication.

2
Centralized Multi-user Key Management

One of the most important challenges for securing group-oriented communications is the issue of key management. As we outlined in the introductory chapter, managing keys in a group-oriented scenario is harder than traditional key management services.

In this chapter, we explore the challenges associated with centralized key management for group-oriented applications. We will begin with an overview of the fundamental limits governing centralized multicast key distribution, and then provide a survey of several approaches that exist in the literature. We then develop a new framework for multicast key management that reduces the communication overhead associated with key management, and show how to best tune this key management scheme to reduce communication overhead.

2.1 Basic Multicast Information Theory

We now provide a summary of information theory results relevant to multicasting. Much of this summary is based upon results that were presented in [3, 4].

First, let $\mathcal{U} = \{u_1, u_2, \cdots, u_n\}$ denote the user set consisting of n users u_j. Associated with each user u_j is a private key K_j that is drawn uniformly from a key space $\mathcal{K}$. Of the n users, only a subset of them will be privileged. We denote the set of private keys associated with privileged members by K_P, and the set of private keys associated with non-privileged users by K_F. For example, if there are $n = 4$ users, and users u_1, u_3 are privileged, then

$K_P = \{K_1, K_3\}$, and $K_F = \{K_2, K_4\}$. There is a secret S that is drawn from a space $\mathcal{S}$ that the group center wishes to transmit to members of the multicast group $\mathcal{U}$. The broadcast message α is a function of the secret S, as well as the private user information of the privileged users, $\alpha = f(S, K_P)$.

It is useful to derive bounds on the size of the broadcast message given the following security constraints:

- The user's secrets K_P and the secret S uniquely determine the broadcast message

$$H(\alpha|S, K_P) = 0. \tag{2.1}$$

- Knowing *only* a user's private key K_j does not decrease the uncertainty of the secret S. That is

$$H(S|K_j) = H(S). \tag{2.2}$$

  In particular, this implies that $H(S|K_P) = H(S)$.

- No uncertainty in the secret remains if both a user's private key K_j and the broadcast message are available.

$$H(S|K_j, \alpha) = 0. \tag{2.3}$$

- The broadcast message does not decrease the uncertainty in a user's private key:

$$H(K_j|\alpha) = H(K_j). \tag{2.4}$$

- The broadcast message alone does not decrease the uncertainty in the secret:

$$H(S|\alpha) = H(S). \tag{2.5}$$

The first results that we present are from Just et al. [3]. In the proofs, we have followed the basic derivations provided in [3].

Lemma 1. *The entropy of the broadcast message α is equal to mutual information between the message and the joint random variable (K_P, S):*

$$H(\alpha) = I(\alpha; K_P, S). \tag{2.6}$$

Proof. We start by applying the chain rule to the mutual information:

$$I(\alpha; K_P, S) = I(\alpha; K_{j_1}) + I(\alpha; K_{j_2}|K_{j_1}) + \cdots$$
$$+ I(\alpha; K_{j_m}|K_{j_1}, K_{j_2}, \cdots, K_{j_{m-1}}) + I(\alpha; S|K_P).$$

Expanding the mutual information terms yields the telescoping sum:

$$I(\alpha; K_P, S) = H(\alpha) - H(\alpha|K_{j_1}) + H(\alpha|K_{j_1}) - H(\alpha|K_{j_1}, K_{j_2}) + \cdots$$
$$+ H(\alpha|K_P) - H(\alpha|K_P, S),$$

which yields

$$I(\alpha; K_P, S) = H(\alpha) - H(\alpha|K_P, S). \tag{2.7}$$

However, $H(\alpha|K_P, S) = 0$, so that

$$I(\alpha; K_P, S) = H(\alpha). \tag{2.8}$$

$\square$

Lemma 2. *Let $D \subset P$ be a subset of privileged members such that $|D| \leq m - 1$, and let K_D be the set of private keys associated with users in D. Let K_i be a private key of a user $u_i \in P - D$. Then for a secret S and a broadcast message α, we have*

$$H(K_i) - H(K_i|\alpha, K_D) \geq H(S). \tag{2.9}$$

Proof. The term $H(K_i, S|\alpha, K_D)$ may be expanded in two different ways:

$$
\begin{aligned}
H(K_i, S|\alpha, K_D) &= H(K_i|\alpha, K_D) + H(S|\alpha, K_D, K_i) & (2.10) \\
&= H(S|\alpha, K_D) + H(K_i|\alpha, K_D, S). & (2.11)
\end{aligned}
$$

Since $H(S|\alpha, K_j) = 0$ for any user u_j in the privileged set P, we have that $H(S|\alpha, K_D, K_i) = H(S|\alpha, K_D) = 0$, and thus

$$H(K_i|\alpha, K_D) = H(K_i|\alpha, K_D, S). \tag{2.12}$$

Observe that since $I(K_i; S|\alpha) = I(S; K_i|\alpha)$ we have

$$
\begin{aligned}
H(K_i|\alpha) - H(K_i|\alpha, S) &= H(S|\alpha) - H(S|\alpha, K_i) \\
H(K_i) - H(K_i|\alpha, S) &= H(S). & (2.13)
\end{aligned}
$$

Since $H(K_i|\alpha, S) \geq H(K_i|\alpha, S, K_D)$, we may apply (2.12) to get $H(K_i|\alpha, S) \geq H(K_i|\alpha, K_D)$. Substituting this result into (2.13) gives $H(K_i) - H(K_i|\alpha, K_D) \geq H(S)$. $\square$

A consequence of this lemma is the fact that each private key K_i will have entropy greater than the entropy of secret, i.e. $H(K_i) \geq H(S)$. We may now put these results together to get a lower bound on the size of the broadcast message given the conditions stated.

Theorem 1. *Suppose that the keys K_j are distributed independently of each other, i.e. $H(K_j, K_i) = H(K_j) + H(K_i)$, and the conditions (2.1)-(2.5) hold, then the following bound on the size of the broadcast message holds:*

$$H(\alpha) \geq mH(S) \tag{2.14}$$

Proof. By Lemma 1 we have

$$
\begin{align}
H(\alpha) &= I(\alpha; K_P, S) \tag{2.15}\\
&= I(\alpha; K_P) + I(\alpha; S|K_P) \tag{2.16}\\
&= I(K_P; \alpha) + I(S; \alpha|K_P) \tag{2.17}\\
&= H(K_P) - H(K_P|\alpha) + H(S) - H(S|\alpha, K_P). \tag{2.18}
\end{align}
$$

Using the fact that $H(S|\alpha, K_P) = 0$ and that

$$
\begin{align}
H(K_P) &= H(K_{j_1}, K_{j_2}, \cdots, K_{j_m}) \tag{2.19}\\
&= H(K_{j_1}) + \cdots + H(K_{j_m}), \tag{2.20}
\end{align}
$$

which follows from the independence of the private keys, we have

$$
H(\alpha) = H(K_{j_1}) + \cdots + H(K_{j_m}) - H(K_P|\alpha) + H(S). \tag{2.21}
$$

Similarly, expanding $H(K_P|\alpha)$ using the chain rule gives

$$
\begin{align}
H(K_P|\alpha) &= H(K_{j_1}|\alpha) + H(K_{j_2}|\alpha, K_{j_1}) + \cdots\\
&\quad + H(K_{j_m}|\alpha, K_{j_1}, \cdots, K_{j_{m-1}}). \tag{2.22}
\end{align}
$$

Upon substitution we get

$$
H(\alpha) = H(K_{j_1}) - H(K_{j_1}|\alpha) + \sum_{i=2}^{m} \left(H(K_{j_i}) - H(K_{j_i}|\alpha, K_{j_1}, \cdots, K_{j_{i-1}}) \right) + H(S). \tag{2.23}
$$

By observing that $H(K_{j_1}|\alpha) = H(K_{j_1})$, and applying Lemma 2 we get the desired result $H(\alpha) \geq mH(S)$. $\square$

   In summary, we have presented two main results from [3] that govern the theoretical underpinnings of multicast key management. The first result that was shown states that the entropy of a user's private information must be greater than the entropy of the secret that is to be distributed to the group. This translates into the security terminology by implying that the bit length of the user's private key should be as large as the bit length of the group secret. It was also shown, under the assumption of independent keys, that the size of the broadcast message must be at least as large the size of the group times the size of the secret that is to be conveyed. This latter result gives a lower bound on the communication requirements for rekeying. In particular, it implies that the best that can be done is a message whose size is linear in the amount of group members unless the key independence assumption is relaxed. As we shall see in the later discussions, the implication of this result is that we must do away with the independence assumption in order to reduce the message size. Currently, the most popular family of multicast key management schemes are those that employ a tree key hierarchy, in which the key information that each user has is not independent of each other.

2.2   Overview of Multicast Key Management

The distribution of identical data to multiple parties using the conventional point-to-point communication paradigm makes inefficient usage of resources. The redundancy in the copies of the data can be exploited in multicast communication by forming a group consisting of users who receive similar data, and sending a single message to all group users [1]. Access control to multicast communications is typically provided by encrypting the data using a key that is shared by all legitimate group members. The shared key, known as the session key (SK), will change with time, depending on the dynamics of group membership as well as the desired level of data protection. Since the key must change, the challenge is in key management–the issues related to the administration and distribution of keying material to multicast group members.

In order to update the session key, a party responsible for distributing the keys, called the group center (GC), must securely communicate with the users to distribute new key material. The GC shares keys, known as key encrypting keys (KEKs), that are used solely for the purpose of updating the session key and other KEKs with group members.

As an example of key management, we present a basic example of a multicast key distribution scheme. Suppose that the multicast group consists of n users and that the group center shares a key encrypting key with each user. Upon a member departure, the previous session key is compromised and a new session key must be given to the remaining group members. The GC encrypts the new session key with each user's key encrypting key and sends the result to that user. Thus, there are $n-1$ encryptions that must be performed, and $n-1$ messages that must be sent on the network. The storage requirement for each user is 2 keys while the GC must store $n+1$ keys. This approach to key distribution has linear communication, computation and GC storage complexity. As n becomes large these complexity parameters make this scheme undesirable, and more scalable key management schemes should be used.

In general, during the design of a multicast application, there are several issues that should be kept in consideration when choosing a key distribution scheme. We now provide an overview of some of these issues.

- **Dynamic nature of group membership:** It is important to efficiently handle members joining and leaving as this necessitates changes in the session key and possibly any intermediate keying information.

- **Ability to prevent member collusion:** No subset of the members should be able to collude and acquire future session keys or other member's key encrypting keys.

- **Scalability of the key distribution scheme:** In many applications the size of the group may be very large and possibly on the order of several million users. The required communication, storage, and computational resources should not become a hindrance to providing the service as the group size increases.

One approach to group key management is provided by the group key management protocol (GKMP) [5]. In this scheme, the GC uses a SK, called a group traffic encrypting key (GTEK) in the GKMP literature, and a group key encrypting key (GKEK). The GC updates the SK by using the GKEK. This allows all group members to be updated using a single encrypted message. A major disadvantage of GKMP, however, is that it is not able to handle member departures, or the compromise of a single member. The compromise of the GKEK means that all future communication is compromised since an adversary can calculate future session keys.

Fiat and Naor [6] present a broadcast key distribution scheme that allows for a single source to transmit a SK to a dynamic subset of privileged users such that no coalition of at most k non-privileged users can acquire the SK. The communication overhead of their scheme is not dependent on the amount of non-privileged members, but instead on the security parameter k and a parameter describing the probability that a coalition of at most k non-privileged users can acquire the SK.

In Section 2.1, we summarized the theoretical work of [3,4] for the distribution of secret information via broadcast messages. These results provide an insight into the communication resources needed to achieve the above goals. In particular, it was shown in Theorem 1 that for a key size of B bits, the message needed to update a group of n users must be at least nB bits to provide *perfect security* in the key distribution. One key result of [3] is that in order to achieve a smaller broadcast size, it is necessary to do away with the constraint that the private information held by each user is mutually independent. Therefore, to reduce the usage of communication resources, the users must share secret information.

One strategy for having users share secret information is to arrange the keys according to a tree structure. The tree based approach to group rekeying was originally presented by Wallner et al. [7], and independently by Wong et al. [8]. In such schemes an a-ary tree of depth $\log_a n$ is used to break the multicast group into hierarchical subgroups. Each member is assigned to a unique leaf of the tree. KEKs are associated with all of the tree nodes, including the root and leaf nodes. A member has knowledge of all KEKs from his leaf to the root node. Thus, some KEKs are shared by multiple users. Adding members to the group amounts to adding more depth to the tree, or adding new branches to the tree [8,9]. Upon member departure the session key and all the internal node KEKs assigned to that member become compromised and must be renewed. Due to the tree structure, the communication overhead is $\mathcal{O}(\log n)$, while the storage for the center is $\mathcal{O}(n)$ and for the receiver is $\mathcal{O}(\log n)$.

Various modifications to the tree scheme have been proposed. In [10], a modification to the scheme of Wallner et al. is presented. By using pseudorandom generators, their scheme reduces the usage of communication resources by a factor of two. Similarly, Balenson et al. [9] were able to reduce the communication requirements by a factor of two using one-way function trees. The security of the Canetti et al. scheme can be rigorously proven, while the security of the approach using one-way function trees is based upon non-standard cryptographic assumptions and has therefor not been rigorously shown. In [11] Canetti et al. examine the tradeoffs between storage and communication requirements, and a modification to the tree-based schemes of [7, 8] is presented that achieves sublinear server-side storage. Further, in [12], it was shown that the optimal key distribution for a group leads to Huffman trees and the average number of keys assigned to a member is related to the entropy of the statistics of the member deletion event.

2.3 Requirements for Centralized Group Key Management

A conditional access system for group communications must be able to cope with the demands of the application. These demands must not only address the security and access requirements of the service provider, but also address the convenience and satisfaction of the client. Below we have listed several functionalities that are desirable in a conditional access system for dynamic group communication scenarios:

1. *The solution should be able to refresh the keys used to protect content.*

   Due to the bulk quantities of data being multicast, it is feasible that session keys may become compromised. Therefore, it is important that there is a means available to refresh the session key and intermediate keying material in order to maintain a desirable level of content protection.

2. *The solution should provide the ability for members to join and depart the service at will, as well as allow the content distributor to easily revoke a member's ability to access content.*

   Unlike unicast communication, the departure of a group member does not imply the termination of the communication link. In addition, upon departing the service, users must be de-registered and prevented from obtaining future multicasts. Similarly, when new members join the service, it is desirable to prevent them from accessing past content. Additionally, situations might arise where the content provider desires to prevent a user from accessing future content.

3. *The solution should be resistant to member collusion.*

   No subset of the members should be able to collude and acquire keying information of non-colluding members.

4. *The solution should provide a means for an end-user to recover from missed rekeying messages.*

   In many application environments, the connection between a client and the server may be severed. For example, in cellular applications, a client might move temporarily through a region of severe fading. Adverse communication conditions and common accidents, such as a system crash, might mean that the client misses several rekeying messages needed to update his key database. Users might also desire to switch from terminal to terminal, with the possibility of not being able to receive communication while moving across terminals. It is important to have a means that allows the client to resume access to the service.

5. *The solution should allow the user to temporarily transfer access rights to another party.*

   In many business scenarios, a client will subscribe to a service where content, such as multimedia or stock quotes, is streamed. Users may wish to transfer their access rights to the data stream to their friends without canceling or transferring their subscription.

6. *The solution should address the issue of resource scalability for scenarios consisting of large privileged groups.*

   In many applications, the size of the group may be very large and possibly on the order of several million users. The required communication, storage, and computational resources should not become a hindrance to providing the service as the group size increases.

Some of these functionalities have been discussed in other tree-based key management schemes. However, many of these objectives are not considered. For the remainder of this chapter we shall present an architecture for the management of keys in a conditional access multicast system that is capable of achieving each of these requirements. The system that we describe makes use of a tree-structured key hierarchy and basic primitive operations to provide a solution that satisfies the above requirements.

Additionally, whereas most of the multicast key management schemes in the literature do not consider the issue of flagging to the user which rekeying messages are intended for them, we provide this important functionality in our message structure and factor this additional overhead into our considerations. We will focus on the usage of communication resources and calculate the amount of communication needed to perform a member join and a member departure operation for different tree degrees and different amount of users. We determine the optimal tree degree for scenarios where member join is most important, member departure is most important, and where both operations are equally important. Then, in order to better study the optimization of the key management scheme, we present a stochastic occupancy model that allows one to study the mean behavior of a key tree under different degrees of occupancy. Additionally, we compare the amount of communication overhead needed in our scheme with the amount of communication overhead that a conventional tree-based rekeying scheme, such as [8], would need to flag users which component of a rekeying message is intended for them.

Looking forward, in Section 2.4 we introduce a method for distributing keys using polynomial interpolation and parametric one-way functions. This basic scheme is used as a building block for a protocol primitive described later in the chapter. Therefore, we present a study of its security and communication features. In Section 2.5 we present some protocol primitives and use these to construct more complex key management operations capable of maintaining the key hierarchy in scenarios with dynamic membership. The size of the messages needed for updating the keys is computed in Section 2.6 and are used to determine the optimal degree of the key distribution tree. Additionally, we examine the computational requirements of the tree-based polynomial interpolation scheme proposed in this chapter.

2.4 Basic Polynomial Interpolation Scheme

The heart of the new multicast key management scheme that we will describe involves the use of a polynomial interpolation algorithm that is capable of reducing the communication overhead needed for key management compared with multicast key management schemes that use messages that are concatenations of individual rekeying messages.

In this section we describe the basic scheme for distributing keys that will be used in the scalable key management protocol of Section 2.5. The basic key distribution scheme that we describe is a modification of the polynomial interpolation scheme of [4]. We have introduced the use of one-way functions and a broadcast seed to protect private user KEKs from compromise and allow private user KEKs to be reused.

We shall use keyed (parametric) one-way functions in our work to provide computational security. A one-way function h is a function from $\mathcal{X} \times \mathcal{Y} \to \mathcal{Z}$ such that given $z = h(x, y)$ and y it is computationally difficult to determine x [13]. Keyed one-way functions, or parametric one-way functions (POWF) are families of one-way functions that are parameterized by the parameter y. Symmetric block ciphers can be used to construct POWFs. Let $x \in \mathcal{X}$ and $y \in \mathcal{Y}$, and consider a symmetric cipher $E_x(y) : \mathcal{Y} \to \mathcal{Y}$ where the subscript denotes the key used in the encryption of the plaintext y. Thus $\mathcal{X}$ is the key space of the cipher E, while $\mathcal{Y}$ is the space of plaintexts and ciphertexts. Define a hash function $f : \mathcal{Y} \to \mathcal{Z}$. Then the function $h(x, y) = f(E_x(y))$ is a POWF parameterized by y since any reasonable cryptosystem can withstand a known-plaintext attack, that is knowledge of $E_x(y)$ and y does not make it easy to determine the key x. Note that it is

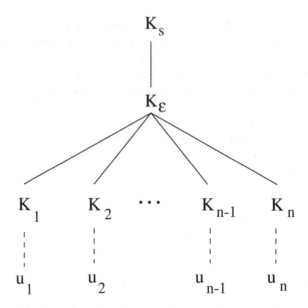

FIGURE 2.1. The basic key distribution scheme used in the polynomial interpolation method.

not necessary that the hash function f have any cryptographic properties as the required cryptographic strength is provided by E. Throughout this chapter we shall assume the existence of parametric one-way functions that map sequences of $2B$ bits into sequences of B bits.

Consider the basic key distribution scheme depicted in Figure 2.1. Each user u_i has a personal B-bit KEK K_i that is known only by the group center and user u_i. Additionally, all of the users share a B-bit root KEK $K_\epsilon(t)$ and a session key $K_s(t)$ that will vary with time t.

Suppose that user u_n decides to depart, then we must renew the keys $K_\epsilon(t-1)$ and $K_s(t-1)$ since they were shared by u_n and the other users. The first step is to send the new $K_\epsilon(t)$ to the remaining users. In the polynomial scheme, each user u_i has the distinct pair $(z_i, K_i) \in Z_p \times Z_p$, where Z_p denotes the integers modulo the prime p. The z_j are public knowledge, and are not considered as part of the secret information that the user must store. Instead, the z_j is any quantity that is used to identify the user, for example a processor id. The GC has made available f, a POWF taking $2B$ bits to B bits. The GC first broadcasts the seed $\mu(t)$ to everyone. Next, the GC associates the following quantity with each user u_j

$$w_j = K_\epsilon(t) + f(K_j, \mu(t)) \pmod{p}. \tag{2.24}$$

The GC generates a degree $n - 2$ polynomial $p(z)$ that interpolates the points (z_j, w_j), i.e. $p(z_j) = w_j$. The GC represents $p(z)$ as

$$p(z) = \sum_{i=0}^{n-2} c_i z^i \quad (\bmod\ p) \qquad (2.25)$$

and transmits the message $\alpha_\epsilon(t) = (c_0, c_1, \cdots, c_{n-2})$ to update $K_\epsilon(t)$. This completes the action needed by the GC to update the root KEK, and the session key is then updated using $K_\epsilon(t)$ by transmitting $\alpha_s(t) = E_{K_\epsilon(t)}$ $(K_s(t))$.

A member u_j can calculate $p(z_j) = w_j$ and $f(K_j, \mu(t))$, and hence can recover $K_\epsilon(t)$.

2.4.1 Resistance to Attack

There are two sources of adversaries for a key management scheme. The first type of adversary is an *external* adversary. This type of adversary is not a member of the service, but receives the encrypted content as well as the rekeying messages. In order for the external adversary to cheat the service, he must mount a successful attack against the rekeying messages in order to acquire the session key, which is needed to decrypt the content. The second type of adversary is an *internal* adversary, who is a member that uses the rekeying messages and his knowledge of his keys to attempt to acquire another user's keys. If an internal adversary can successfully acquire another user's keys, he may cancel his membership to the service, and use the compromised keys belonging to another user to enjoy the service without having to pay.

In the polynomial scheme, an external adversary receives α_ϵ as well as $\alpha_s(t)$. In order for the adversary to acquire the SK, he must mount a successful attack against the cipher used in forming the message $\alpha_s(t)$. Careful selection of a strong cipher algorithm that has received serious study, such as Rijndael [14], will make a successful attack of the SK rekeying message unlikely. Even should a successful attack of the SK rekeying message take place, a future update of the SK would require a subsequent successful attack of the SK rekeying message, which is equally unlikely. Hence, a successful attack against the SK rekeying message would only be a short-lived victory for a pirate.

A second method for acquiring the session key is to attack the message α_ϵ. Given the message $\alpha_\epsilon(t)$, and knowledge of a z_j, it is possible that an adversary may calculate w_j. However, the adversary must either determine $K_\epsilon(t)$ or a user's $f(K_j, \mu(t))$ given $w_j = K_\epsilon(t) + f(K_j, \mu(t)) \pmod{p}$. The modulo operation makes w_j independent of either $K_\epsilon(t)$ or $f(K_j, \mu(t))$. Should an external adversary successfully attack $K_\epsilon(t)$, then he may acquire the session key. However, upon the next update of the session key, he must make another successful attack upon the root KEK.

The only method for an external adversary to be able to repeatedly acquire the SK is to mount a successful attack on a user's personal key K_j. This requires successful determination of $f(K_j, \mu(t))$ given w_j, which requires searching a space of order p possibilities, and then successfully attacking the one-way function to acquire K_j. The strength of the one-way function should be as strong as the strength of the encryption used to protect the SK rekeying message.

We now discuss the susceptibility of the original polynomial scheme of [4] to *internal* attacks. In the discussion that follows, we refer the reader to Section 3.1 of [4]. For simplicity, we shall assume that the same key K is being distributed to all of the users. Observe that since the z_j-coordinates are public knowledge, an internal adversary may calculate w_j by evaluating the interpolating polynomial at z_j. With knowledge of w_j, the adversary may use his knowledge of K to determine user u_j's private information. Thus, the polynomial scheme of [4] does not protect the private information of each user, and hence cannot be used more than once. If both the z_j coordinate and the personal key K_j are kept secret, then an adversary's task is to search Z_p for any of the n user's z_j coordinate. This is more difficult for an adversary to attack, but also requires both the server and the clients to store twice as much secret information.

As we shall describe in Section 2.6.3, we chose to pursue a different approach to ensuring the sanctity of each user's private information in order to reduce the communication overhead in our protocol. An inside adversary u_i who desires to calculate another user's key information K_j can calculate $p(z_j) = w_j$, and therefore can calculate $f(K_j, \mu(t)) = w_j - K_\epsilon(t)$ (mod p). However, it is difficult for him/her to calculate K_j given $\mu(t)$ and $f(K_j, \mu(t))$ since f is a parametric one-way function. Additionally, should two or more users collude, their shared information does not provide any advantage in acquiring another user's K_j.

2.4.2 Anonymity Reduces Communication Overhead

The above scheme is used in constructing a protocol primitive in the following section. In the protocol primitive, there is a parent key K_ϵ and a handful of sibling keys K_j that are used to update the parent key. Unlike the example described above, application of the protocol primitive might not use all of the sibling keys to update the parent key. This scenario might occur when the GC knows that a sibling key has become compromised or invalidated.

Suppose that there are a possible sibling keys and that m of those sibling keys are used to update the parent key. In a conventional key distribution scheme, such as [8], the update to the parent key is performed by a rekeying message of the form

$$\alpha = \{E_{K_{j_1}}(K_\epsilon) \| E_{K_{j_2}}(K_\epsilon) \| \cdots \| E_{K_{j_m}}(K_\epsilon)\} \qquad (2.26)$$

where j_k denotes the sequence representing the m sibling keys used in updating parent key, and $\|$ denotes message concatenation. In addition to the rekeying message, it is necessary to transmit the amount m of children keys, and the user ID message $\{j_1, j_2, \cdots, j_m\}$, which specifies which portion of the rekeying message a user needs in order to determine the new session key.

The transmission of the user ID message in the conventional scheme reveals which sibling keys are still valid. However, it requires that $\lceil \log_2 a \rceil$ bits to represent m and $m\lceil \log_2 a \rceil$ bits to represent $\{j_1, j_2, \cdots, j_m\}$. The total communication overhead of the conventional scheme is thus $(m + 1)\lceil \log_2 a \rceil$ bits.

The polynomial interpolation scheme creates a composite message that does not require any user ID message, but instead requires the broadcast of the seed $\mu(t)$. The polynomial scheme defines the rekeying message as the output of a function $PolyInt$ which returns the coefficients of the interpolating polynomial, thus

$$\alpha = PolyInt(K, \{z_{j_1}, z_{j_2}, \cdots, z_{j_m}\}, \{K_{j_1}, K_{j_2}, \cdots, K_{j_m}\}, \mu(t)). \quad (2.27)$$

The input to $PolyInt$ is the key K that is to be distributed, the set of valid non-secret ID parameters $\{z_{j_1}, z_{j_2}, \cdots, z_{j_m}\}$, the broadcast seed $\mu(t)$, and the set of valid sibling keys $\{K_{j_1}, K_{j_2}, \cdots, K_{j_m}\}$. Given a valid sibling key and the seed $\mu(t)$, the new parent key can be determined. On the other hand, an invalid sibling key is unable to determine the new parent key.

If the prime p used in the polynomial scheme has the same bit length as the output of one of the encryptions E_K, then the message size of the polynomial scheme will be the same as the rekeying message of the conventional scheme. If B_μ is the bit length of the broadcast seed, then a measure of comparison between the conventional scheme and the polynomial scheme is the difference $(m + 1)\lceil \log_2 a \rceil - B_\mu$. For a single sibling update of the parent node, this difference might favor the conventional approach. The advantage of the polynomial scheme becomes more pronounced when used in a multi-level tree as in Section 2.5.

2.5   Extending to a Scalable Protocol

In the previous section we described the basic scheme for distributing keys during member departures. The basic polynomial interpolation scheme had linear communication requirements during member departures. We now describe a scalable protocol that provides renewal of security levels, handles membership changes, provides a mechanism for reinserting valid members, and allows for the transferal of access rights.

In order to achieve improved scalability, we use a tree-based key hierarchy as depicted in Figure 2.2. In general, the tree can be an a-degree tree.

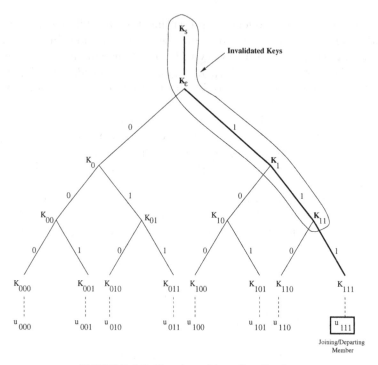

FIGURE 2.2. Tree-based key distribution.

Attached to the tree above the root node is the session key K_s. Each node of the tree is assigned a KEK which is indexed by the path leading to itself. Additionally, each node has a non-secret ID variable z_σ which is used as a non-secret parameter for the $PolyInt$ function. The symbol ϵ is used to denote the root node. Each user is assigned to a leaf of the tree and is given the KEKs of the nodes from the leaf to the root node. Additionally, all users share the session key K_s. For example, user u_{111} is given the keys K_{111}, K_{11}, K_1, K_ϵ, and K_s.

In the protocol that follows, the GC transmits messages to the users via a broadcast channel. It is assumed that each user has an *upstream* channel with minimal bandwidth that is available to convey messages to the GC, such as informing the GC of the intent to depart the service.

The messages that the GC broadcasts to the users must have a standardized structure that is known to all receivers. There are two basic message formats as depicted in Figure 2.3. The first contains three components while the second has five components. The function $B()$ is used to denote the bit length of its operand, thus $B(\sigma)$ is the amount of bits needed to represent σ. The variable Operation ID flags the user which protocol primitive is about to be performed. Only five primitive operations are used, and we may therefore represent Operation ID using a 3 bit string. Table 2.1 maps the primitive operations with their corresponding ID bit string.

TABLE 2.1. Mapping between primitive operations and their corresponding ID bit string.

bit ID	primitive
000	Primitive-1
001	Primitive-2
010	Primitive-3
011	Primitive-4
100	Primitive-5

In the discussion that follows, we assume that the tree has degree a, and that there are L levels to the tree. The amount of multicast group members n is limited by the amount of leaf nodes on the tree. Thus $n \leq a^L$.

2.5.1 Basic Protocol Primitives

We have identified five basic operations needed in building a system that allows for the update and renewal of the key hierarchy. We now describe each case.

1. **Primitive-1(Update SK):** This basic operation uses the current root KEK K_ϵ to update the session key via the rekeying message

$$\alpha = E_{K_\epsilon(t)}[K_s(t)] \tag{2.28}$$

   The message format is depicted in Figure 2.3(a). We assume that the maximum size that α can be is 256 bits, and we therefore need 8 bits to represent $B(\alpha)$. This choice of bit length for α would allow for the use of encryption algorithms with a key size of up to 256 bits.

2. **Primitive-2(Transmit Seed):** The broadcast seed is used in the polynomial scheme to provide protection of secret information. Additionally, it plays a role in reducing the communication overhead associated with flagging the users which part of the message is intended for them. The broadcast of the seed $\mu(t)$ does not require encryption to protect it. The message format for the transmission of the broadcast seed is depicted in Figure 2.3(a). Here $\alpha = \mu(t)$, and $B(\alpha)$ is the amount of bits needed to represent $\mu(t)$. Again, we assume that the maximum size of α is 256 bits, and that 8 bits are used to represent $B(\alpha)$.

3. **Primitive-3(Self Update):** It is often necessary for a node, indexed by the a-ary symbol σ, to have its associated key updated using the key at the previous time instant. Thus we will go from $K_\sigma(t-1)$ to $K_\sigma(t)$ by the following message

$$\alpha = E_{K_\sigma(t-1)}[K_\sigma(t)]. \tag{2.29}$$

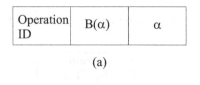

(a)

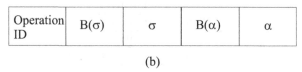

(b)

FIGURE 2.3. The two message structures used in the protocol primitives.

In this case, we need to flag the receivers which node is being updated. This requires the transmission of the a-ary representation of the node, as well as the amount of bits needed to represent the node. This is depicted in Figure 2.3(b) by the $B(\sigma)$ and σ components of the message. The rest of the message contains the bit length of the message α and the actual rekeying message α. Since the maximum depth of the tree that needs to be represented is $L-1$ and the tree is an a degree tree, the maximum amount of bits needed to represent σ is $\lceil \log_2 a \rceil (L-1) + 1$, where the addition of 1 bit was included to account for the need to represent the empty string ϵ as a possible choice for ϵ. In order to represent $B(\sigma)$, we use $\lceil \log_2(\lceil \log_2 a \rceil (L-1) + 1) \rceil$ bits. The maximum bit length for α is 256 bits, and 8 bits are used to represent $B(\alpha)$.

4. **Primitive-4(Update Parent):** It is also necessary for the children nodes to update the key of their parent nodes. If σ is the symbol representing the parent node to be updated, then the message

$$\alpha = PolyInt(K_\sigma(t), \{z_{Child(\sigma)}(t)\}, \{K_{Child(\sigma)}(t)\}, \mu(t)) \qquad (2.30)$$

is used. Here we have defined the function $Child(\sigma)$ to denote the set of valid children nodes of σ. For example, if we have a binary tree and $\sigma = 00$, and both children nodes are valid, then $Child(\sigma) = \{000, 001\}$. Thus, the message α uses the keys of valid children nodes to update $K_\sigma(t)$. Observe that this message requires that $\mu(t)$ has already been broadcast using Primitive-2, or that the choice of $\mu(t)$ is implicitly known. The message form is depicted in Figure 2.3(b), where again we transfer the bit length of σ and the actual symbol σ to the recipients, followed by the bit length of α and the rekeying message α. We use the same bit allocation for σ and $B(\sigma)$ as in Primitive-3. However, the maximum length for α is aB_{KEK}, and we therefore need $\lceil \log_2 aB_{KEK} \rceil$ bits to represent $B(\alpha)$.

5. **Primitive-5(Reaffirming Parent):** In some operations, it is useful to have a sibling node reaffirm the value of a parent node's key. We

define a function $Par(\sigma)$ to denote the symbol corresponding to the parent of the node indexed by σ. To reaffirm the value of a parent node's key, we transmit the message

$$\alpha = E_{K_\sigma(t)}[K_{Par(\sigma)}(t)]. \tag{2.31}$$

The message form is depicted in Figure 2.3(b), and follows the same structure as used in Primitive-3.

2.5.2 Advanced Protocol Operations

We now describe more advanced protocol operations that can be constructed using the primitive operations described above. In particular, we focus on the operations of an addition to the membership, a deletion of a user from the membership, the reinsertion of a member into the system, and the transferal of access rights from one user to a new user.

Before we proceed, we present a few comments about how the primitive operations can be used to perform periodic renewal of keying material. Primitive-1 provides a method for performing periodic refreshing of the session key. Refreshing the session key is important in secure communication. As a session key is used, more information is released to an adversary, which increases the chance that a SK will be compromised. Periodic renewal of the session key is required in order to maintain a desired level of content protection, and can localize the effects of a session key compromise to a short period of data. Since the amount of data encrypted using KEKs is usually much smaller than the amount of data encrypted by a session key, it is not necessary to refresh KEKs as often. However, the periodic renewal of a KEK can be performed using Primitive-3.

Member Join

In many applications, such as pay-per-view broadcasts and video conferences, the group membership will be dynamic. It is important to be able to add new members to any group in a manner that does not allow new members to have access to previous data. In a pay-per-view system, this amounts to ensuring that members can only watch what they pay for, while in a corporate video conference there might be sensitive material that is not appropriate for new members to know.

Suppose that a new user contacts the service desiring to become a group member. The new client sends the GC a message detailing the client's credentials, such as identity information, billing information, and public key parameters that the GC may use to communicate with the new client. Mutual authentication between the new client and the GC should be performed. A public key infrastructure, such as X.509 certificates [15], may be used for this purpose. Upon verification of the new user's information, the GC assigns the client to an empty leaf of the key tree. For simplicity of presentation, we assume that the tree has empty slots. If the tree is already full, then the user may either be turned away, or an additional layer must be added to the tree using a separate operation. The GC then issues the new client his keys via a communication separate from the communications sent to the current group members, as well as informing the new user the time at which those keys will become valid.

Meanwhile, the GC updates the current members of the multicast group. Suppose that the GC plans on inserting the new member into the leaf node indexed by the symbol ω. Then the SK as well as the KEKs on the path from the parent node of ω to the root node ϵ must be renewed. The following algorithm describes how this procedure can be accomplished using the protocol primitives. We use the notation $Par^j(\omega)$ to denote the parent function applied j times to ω. Thus $Par^2(\omega)$ is the *grandparent* of ω.

---

for $j = 1 : L$ **do**
    $\sigma = Par^j(\omega)$;
    Update $K_\sigma(t-1) \rightarrow K_\sigma(t)$ using Primitive-3 ;
end
Update SK using Primitive-1 ;

Member Departure

Members will also wish to depart the service, and must be prevented from accessing future communication. Assume that user u_ω contacts the GC wishing to depart the service. Upon authenticating the user's identity, the procedure that the GC enacts to remove member u_ω and update the keys of the remaining members is

---

Generate random $\mu(t)$;
Broadcast $\mu(t)$ using Primitive-2 ;
for $j = 1 : L$ **do**
    $\sigma = Par^j(\omega)$;
    Determine valid children of σ: $Child(\sigma)$;
    Update $K_\sigma(t-1) \rightarrow K_\sigma(t)$ using Primitive-4 ;
end
Update SK using Primitive-1 ;

Member Reinsertion

It might often occur that a valid member, denoted by index ω, misses the rekeying messages needed to update the key hierarchy. The client must notify the GC that he missed rekeying messages using an upstream (client to server) channel. Upon verification of the user's identity, the GC performs the member reinsertion operation, which sends the new user the specific keys he needs to be able to resume the service.

If the service provider has a downstream (server to client) channel available to communicate with the user, then service provider may use this channel to send the needed keys by encrypting them with the user's personal key K_ω. In many scenarios, however, after the initial contact with the service provider, the client has a low-bandwidth channel for upstream communication, and only the broadcast channel available for downstream communication. In these cases, although only a single user needs the rekeying messages, the rekeying messages must be multicast. Since this user has a valid private key K_ω, the GC can start with this key to provide $K_{Par(\omega)}(t)$ to the user. We can then proceed up the tree, using the sibling key to convey the current status of the parent key. The procedure for this operation is as follows:

---

for $j = 1 : L$ **do**
   $\sigma = Par^j(\omega)$;
   Convey parent key $K_\sigma(t)$ to siblings using Primitive-5 ;
end
Convey current SK using Primitive-1 ;

An added bonus of using the sibling key to convey the current status of the parent key is that other users may observe these rekeying messages to reaffirm the validity of some of their keys.

Transferal of Rights

Suppose that user u_ω wishes to give his rights to another user who is not currently a member. We will denote this new user by u_{ω_B} to indicate that he will take over the keys on the path from ω to the root node. For the purpose of calculating parent and sibling relationships, ω and ω_B are identical, thus $Par(\omega) = Par(\omega_B)$.

In order to transfer access rights, both users must contact the GC, who performs an authentication procedure to verify that the transferal is legitimate. Then, using a secure channel, the GC gives to user u_{ω_B} its own personal key K_{ω_B}. One method for creating a secure channel is to use public key cryptography. K_{ω_B} replaces K_ω on the key tree. All of the keys that belonged to u_ω must be changed to prevent u_ω from accessing content that he has given up the right to access. The procedure for transferring access rights is as follows:

We observe that the algorithm for transferring rights is nearly identical with the algorithm for removing a member from a group. The difference lies in the fact that user u_{ω_B} is considered a valid user, and hence is a valid child of its parent.

Generate random $\mu(t)$;
Broadcast $\mu(t)$ using Primitive-2 ;
for $j = 1 : L$ **do**
   $\sigma = Par^j(\omega_B)$;
   Determine valid children of σ: $Child(\sigma)$;
   Update $K_\sigma(t-1) \rightarrow K_\sigma(t)$ using Primitive-4 ;
end
Update SK using Primitive-1 ;

The procedure for user u_ω to reclaim his access privileges is similar. This time, only user u_ω is required to contact the GC requesting that he regain his access privileges. The GC performs an authentication procedure to guarantee that the identity of u_ω is truthful, and then replaces K_{ω_B} with K_ω. The KEKs and SK are changed according to the above algorithm, with ω replacing ω_B.

2.6   Architectural Considerations

2.6.1   Optimization of Tree Degree for Communication

The amount of communication that a rekeying protocol requires affects the speed at which the rekeying scheme can handle membership changes. It is therefore important to minimize the size of the communication used by the key management scheme. In particular, since the two most important operations performed by a multicast key management protocol are membership joins and membership departures, we shall focus on optimizing the tree degree for these two operations.

In what follows, we present a worst-case analysis of the communication requirements for member join and member departure operations. It is observed that member join and member departure operations lead to conflicting optimality criteria. Since a real system will have to cope with both member joins and member departures, we jointly consider the departure and join operations, and present optimization results when both member join and departure operations are equally weighted.

We refer the reader to the protocol descriptions as well as the message structure in Figure 2.3. We shall denote the degree of the tree by a, and the number of levels in the tree by L. B_{SK} shall denote the bit length of session key, B_{KEK} shall denote the bit length of the key encrypting keys, and B_μ the bit length of the broadcast seed $\mu(t)$.

Worst-Case Analysis

It is easy to see that, for a given tree, the scenario that produces the most communication for the member join operation occurs when one node on each level from the root to level $L - 1$ must be updated. In this case, all of the KEKs on the path from one user to the root must be refreshed. We now calculate the amount of communication needed to update the tree for this worst-case scenario.

The member join operation consists of two types of operations: updating the KEKs, and updating the SK. In order to update the KEKs, we use Primitive-3 L times. Each step of the loop must send the quintuple (operation ID, bit length of update node $B(\sigma)$, node ID σ, bit length of the update message $B(\alpha)$, update message α). The symbol σ starts near the bottom of the tree, and through application of the Parent function moves toward the root of the tree.

In order to represent the symbol σ during the jth iteration of the loop, we need to convert from base a to base 2 and hence $B(\sigma) = \lceil \log_2 a \rceil (L - j) + 1$ bits. In addition, we must send $B(\sigma)$, which requires

$$\lceil \log_2 (\lceil \log_2 a \rceil (L - 1) + 1) \rceil$$

bits. Here the addition of 1 was to allow for the need to represent the empty string ϵ as a possible choice for σ. Similarly, in each stage of the loop the rekeying message α has bit length $B(\alpha) = B_{KEK}$ and since we have fixed the maximum key length to be 256 bits, we require 8 bits to represent $B(\alpha)$. The update to the session key requires sending the ID flag, $B(\alpha)$ and α. Therefore, the amount of bits needed to update the session key is $3 + 8 + B_{SK}$. The total amount of bits needed to update the key tree during a member join is

$$C_{MJ} = \left(\sum_{j=1}^{L} \left[3 + \lceil \log_2 (\lceil \log_2 a \rceil (L - 1) + 1) \rceil + \lceil \log_2 a \rceil (j - 1) + 9 + B_{KEK} \right] \right)$$
$$+ 3 + 8 + B_{SK}.$$

The amount of communication needed in the member departure case can be similarly calculated. The main difference between member join and member departure is that there are three operations: the broadcasting of $\mu(t)$, updating the KEKs, and updating the SK. The most communication occurs when $a - 1$ nodes on level L must be used to update the key on level $L - 1$, and a nodes are used to refresh each of the remaining KEKs on the path from the departing member to the root node. After appropriately expanding and gathering terms, the communication for the member departure can be found to be

$$C_{MD} = 22 + B_{SK} + B_\mu + (La - 1)B_{KEK} + (L)$$
$$\left(4 + \lceil \log_2 a B_{KEK} \rceil + \lceil \log_2 (\lceil \log_2 a \rceil (L - 1) + 1) \rceil + \frac{(L - 1)}{2} \lceil \log_2 a \rceil \right).$$

We calculated the worst-case amount of communication required to update an a-degree key tree as a function of the number of users n with the amount of tree levels set to $L = \lceil \log_a n \rceil$. In our calculations, we chose $B_{SK} = B_{KEK} = B_\mu = 64$ bits. We chose to use 64 bits as the key size since such a key length can provide strong levels of security when used with some ciphers, such as RC5 [13]. The amount of communication required for different choices of the degree of the tree a during a member join is depicted in Figure 2.4(a). This figure shows the general trend that less communication is required during member join operations if we use a higher degree tree. On the other hand, Figure 2.4(b) shows the amount of communication needed during the worst case of a member departure operation. In this case, the larger tree degrees are definitely not advantageous. It is also evident that a binary tree is not optimal when considering member departure. In fact, the values of $a = 3$ and $a = 4$ appear to be the best choice, with optimal choice fluctuating depending on n.

Joint Departure-Join Optimization

In some application scenarios the key tree might start out relatively empty, and the amount of member join operations would be greater than the amount of member departure operations. In this case, the membership grows towards the tree capacity, and the communication required for the member join operation is more critical than the communication for member departure. On the other hand, some scenarios might start out with a nearly full key tree, and the member departure operation would outweigh the member join operation.

We therefore would like a communication measure that runs the gamut between the two extremes of just the member join communication, and just the member departure communication. This can be accomplished by considering the convex combination of C_{MJ} and C_{MD}.

Let λ denote the probability of a member departure operation, and assume that $1 - \lambda$ is the probability of a member join operation, then the combined communication measure C_C given by

$$C_C = \lambda C_{MD} + (1 - \lambda)C_{MJ} \tag{2.32}$$

weights the member departure and member join operations according to their likelihood. For example, when $\lambda = 0$ the emphasis is entirely placed on the member join operation, while $\lambda = 1$ corresponds to when the emphasis entirely placed on the member departure operation. The case of $\lambda = 0.5$ corresponds to equal emphasis on the two operations, which is depicted in Figure 2.5. From this figure, we see that the choice of $a = 4$ stands out as the best choice for $n > 10000$ when equally weighting the member join and member departure operation.

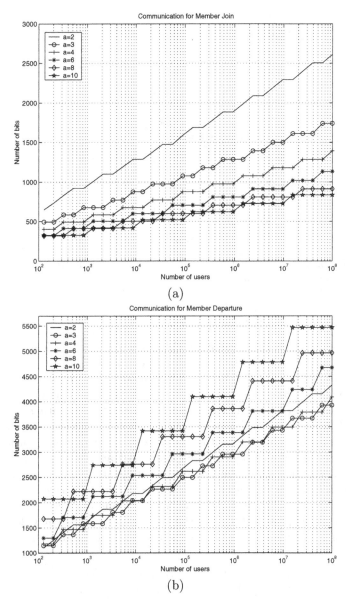

FIGURE 2.4. (a) The amount of communication C_{MJ} required during member join operations for different tree degrees a and different amounts of users n. (b) The worst case amount of communication C_{MD} required during member departure operations for different tree degrees a and different amounts of users n.

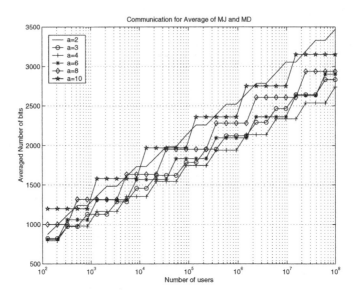

FIGURE 2.5. The average of C_{MD} and C_{MJ} for different tree degrees a and different amounts of users n.

2.6.2   Binomial Occupancy Model

Since it is very difficult to calculate the amount of communication needed during membership changes when a specific amount of users n are placed on the tree, we have devised a stochastic model that allows one to study the behavior of the system when there are varying amounts of occupancy. We assume that the leaf nodes of the a-degree key tree with L levels are occupied according to i.i.d. Bernoulli distributions with a probability of occupancy q_L. This implies that the occupancy n is modeled according to a binomial distribution with mean occupancy $q_L a^L$ and variance $q_L(1-q_L)a^L$. Hence, when q_L is higher, the tree is on average at higher occupancy.

We first calculate the average amount of communication required for member join when the probability of a node being occupied is q_L. Let τ_a denote the a-ary representation of the joining member. We may denote the siblings of τ_a by $\tau_1, \tau_2, \cdots, \tau_{a-1}$. Define the random variable Z_{L-1} as

$$Z_{L-1} = \begin{cases} 1 & \text{if any } \tau_k \text{ is occupied} \\ 0 & \text{if no } \tau_k \text{ are occupied} \end{cases}.$$

Since the τ_k are occupied with a probability of q_L, we have $P(Z_{L-1} = 1) = 1 - (1 - q_L)^{a-1}$, and the expected value of Z_{L-1} is given by $E(Z_{L-1}) = 1 - (1 - q_L)^{a-1}$.

We may perform a similar procedure for the other levels. We denote the j-siblings as those nodes τ such that $Par^j(\tau) = Par^j(\tau_a)$. For level $L - j$, we may define the random variable Z_{L-j} as

$$Z_{L-j} = \begin{cases} 1 & \text{if any } j\text{-sibling node of } \tau_a \text{ is occupied} \\ 0 & \text{if no } j\text{-sibling nodes of } \tau_a \text{ are occupied} \end{cases}.$$

In this case, $P(Z_{L-j} = 1) = 1 - (1 - q_L)^{a^j - 1}$, and the expected value of Z_{L-j} is given by $E(Z_{L-j}) = 1 - (1 - q_L)^{a^j - 1}$.

The average communication requirements for member join can be derived as

$$\overline{C_{MJ}} = \left(\sum_{j=1}^{L} (1 - (1 - q_L)^{a^j - 1}) \left[12 + \lceil \log_2(\lceil \log_2 a \rceil (L - 1) + 1) \rceil \right. \right.$$

$$\left. \left. + \lceil \log_2 a \rceil (L - j) + B_{KEK} \right] \right) + 11 + B_{SK}.$$

We now apply the model to calculating the average amount of communication needed during member departure. Again suppose that the departing member is indexed by the a-ary symbol τ_a. Label the siblings of τ_a by $\tau_1, \tau_2, \cdots, \tau_{a-1}$, and define the random variable X_k by

$$X_k = \begin{cases} 1 & \text{if } \tau_k \text{ is occupied} \\ 0 & \text{if } \tau_k \text{ is not occupied} \end{cases}.$$

Let us define $Y_L = \sum_{k=1}^{a-1} X_k$, which is the random variable corresponding to the amount of occupied sibling nodes of τ_a at level L. The probability that i sibling leafs at level L are occupied is given by

$$P(Y_L = i) = \binom{a - 1}{i} q_L^i (1 - q_L)^{a - 1 - i}. \tag{2.33}$$

Y_L is thus a binomial random variable with expected value $E(Y_L) = (a - 1)q_L$. Hence, the average number of nodes to be updated at level L is $(a - 1)q_L$.

At level $L - 1$, we know that the parent node of the departing member will automatically be used in updating the next higher level. Since the probability of a node at level L being occupied is q_L, the probability that a node on level $L - 1$, other than $Par(\tau_a)$, being occupied is

$$q_{L-1} = 1 - (1 - q_L)^a. \tag{2.34}$$

This time, we may denote the siblings of $Par(\tau_a)$ by $\tau_1, \tau_2, \cdots, \tau_{a-1}$. Again, we define the random variable X_k by

$$X_k = \begin{cases} 1 & \text{if } \tau_k \text{ is occupied} \\ 0 & \text{if } \tau_k \text{ is not occupied} \end{cases}.$$

We now define the random variable Y_{L-1} to be the amount of sibling nodes of $Par(\tau_a)$ that are occupied, and we find that $E(Y_{L-1}) = (a - 1)q_{L-1}$. Since we must also include $Par(\tau_a)$ in the updating we must add one. Thus, the expected number of nodes on level $L - 1$ that must be updated is $1 + (a - 1)q_{L-1}$. We may similarly perform this calculation for level j, where $q_j = 1 - (1 - q_{j+1})^a$, and the expected number of nodes on level j to be updated is $1 + (a - 1)q_j$.

In order to calculate the average amount of communication for the member departure operations, we must consider both the expected amount of communication associated with the overhead and the payload of the message. The average communication for the overhead consists of the amount of communication needed to send the operation id, the node id, and the bit length of the update message. This calculation can be done using the expected value of Z_{L-j}. The average communication for the payload is calculated using the expected number of nodes on level j to be updated. The average amount of communication for n users on an a-degree tree with L levels is therefore given by

$$
\overline{C_{MD}} = 22 + B_\mu + B_{SK} + q_L(a-1)B_{KEK} + \left(\sum_{j=1}^{L-1} (1 + (a-1)q_j) B_{KEK} \right)
$$

$$
+ \left(\sum_{j=1}^{L} \left(1 - (1 - q_L)^{a^j - 1} \right) \left(4 + \lceil \log_2(\lceil \log_2 a \rceil (L-1) + 1) \rceil \right. \right.
$$

$$
\left. \left. + \lceil \log_2 a \rceil (L - j) + \lceil \log_2 a B_{KEK} \rceil \right) \right).
$$

We calculated the mean message size for member join and member departure operations as parameterized by q when the tree degree is $a = 4$ and there are 6, 8 and 10 levels. The key sizes were chosen to be $B_{SK} = B_{KEK} = B_\mu = 64$ bits. In Figure 2.6, we have indicated the mean communication as a function of q. One can see that the expected communication rapidly increases as the probability q becomes slightly greater than 0. In the member join operation, the communication levels off to a flat plateau as the probability of occupancy increases. For the member departure operation, the mean communication also increases rapidly for $q < 0.1$, but then grows less dramatically for higher q. From these two curves, we can infer that a key tree which is roughly half occupied does not have considerably different communication requirements than the worst-case communication requirements, which occur when $q = 1$. This supports our use of the worst-case scenarios for optimizing the tree degree.

2.6.3 Communication Overhead

Earlier we mentioned that one motivation for using the broadcast seed is that it reduces the amount of communication overhead associated with

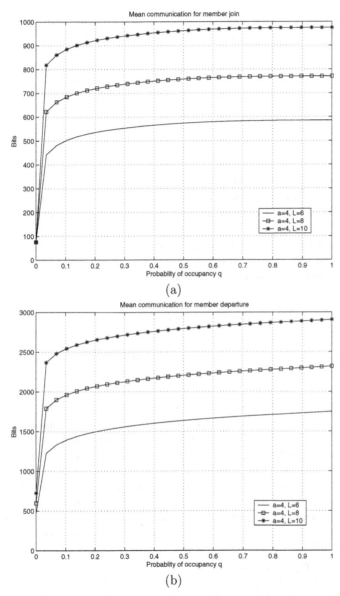

FIGURE 2.6. The expected amount of communication for a degree 4 tree with 6, 8, and 10 levels as a function of the probability q that a leaf node is occupied. (a) Member Join, (b) Member Departure.

notifying to the users which rekeying messages are intended for them during member departures. We now explore this concept in the framework of a tree-based scheme.

Consider an a degree tree with n users. In a general tree-based scheme, when a user departs, all of the keys on the path from the departing member's leaf to the root key must be updated. To update a key associated with a particular node σ, we must determine the keys associated with populated children nodes. These keys are then used to encrypt the update, and the rekeying message is then of the form:

$$\alpha = \{E_{K_{j_1}}(K_\sigma) \| E_{K_{j_2}}(K_\sigma) \| \cdots \| E_{K_{j_m}}(K_\sigma)\}. \tag{2.35}$$

Here we have used the sequence $\{j_k\}$ to denote index the symbols of the valid children nodes. In addition to sending the rekeying message, it is necessary to send the number of valid children nodes m, and the sequence $\{j_1, j_2, \cdots, j_m\}$.

The worst case scenario for communication overhead in updating a tree is when a of the children nodes are used to update each parent node. In this case, the communication overhead required is

$$C_O = (a+1) \lceil \log_2 a \rceil \lceil \log_a n \rceil. \tag{2.36}$$

This equation is obtained by considering both the communication needed to send the amount of valid children nodes, and the symbols for each valid child node.

This amount of communication overhead was calculated for different group sizes n and different tree degrees a. The resulting amount overhead is depicted in Figure 2.7. In this figure we have also drawn a baseline corresponding to $B_\mu = 64$ bits, which is the amount of communication overhead required if one uses the Member Departure protocol of Section 2.5. Examining the case of $a = 4$, which corresponds to the optimal value of the tree-degree as previously determined, shows that for values of $n > 10000$, the Member Departure protocol described in this chapter requires less communication overhead in the worst case scenario. Additionally, observe that if we use a higher degree tree, which is better suited to scenarios where more users are joining than departing, then the efficiency of the Member Departure protocol is even more pronounced.

The use of a broadcast seed can gain further improvement if we choose to use $\mu(t) = K_s(t-1)$. In this case, the broadcast seed does not have to be sent since it is known by the remaining users. Therefore, there is no communication overhead associated with updating during member departure, and we may consider the baseline at $B_\mu = 0$. In this case, the benefits of using a broadcast scheme becomes even more pronounced.

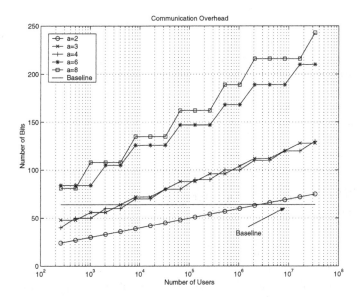

FIGURE 2.7. The worst-case member departure communication overhead required in a conventional tree-based rekeying for different tree degrees versus the baseline communication required when using the polynomial interpolation scheme. The baseline communication corresponds to $B_\mu = 64$ bits.

2.6.4 Computational Complexity

We have seen that one advantage of broadcast schemes is that they reduce the amount of communication overhead associated with sending flagging messages. It should be apparent that a message form like equation (2.26) takes less computation to form than a message form like equation (2.27) assuming that calculating $E_K(K_\sigma)$ has comparable computation as $f(K_\sigma, \mu(t))$. Hence, to rekey using our message form requires more computation than when using a conventional rekeying message structure.

In the scheme we have described in this chapter, we have L levels of KEKs to update. At each level of the tree we must calculate the coefficients of a degree $a - 1$ interpolating polynomial, except at the bottom level where we must calculate the coefficients of a degree $a - 2$ polynomial.

In order to calculate the coefficients of a s-degree interpolating polynomial, we use the Newton form of the interpolating polynomial [16]. Algorithm 1 is a modification of the polynomial interpolation algorithm of [17], which can be used to determine the coefficients β_j of the s-degree polynomial that interpolates the points $(z_j, g_j) \in Z_p \times Z_p$, where $j \in \{0, 1, \cdots, s\}$. The algorithm writes the β_j values into the input array values g_j.

This algorithm requires addition, multiplication, inversion, and modulo operations to take place modulo p. The most intensive operation of these is that of inverting a number. Assume that the prime p is chosen to have B bits, then the amount of bits operations needed to calculate the inverse

---

for $k{=}0{:}s{-}1$ **do**
    for $j{=}s{:}{-}1{:}k{+}1$ **do**
        $g(j) = (g(j) - g(j-1))(z(j) - z(j-k-1))^{-1}$ $(\bmod\ p)$
    end
end
for $k{=}s{-}1{:}{-}1{:}0$ **do**
    for $j{=}k{:}s{-}1$ **do**
        $g(j) = g(j) - g(j+1)z(k)$ $(\bmod\ p)$
    end
end

Algorithm 1: Algorithm for determining the coefficients of an interpolating polynomial.

of a number modulo p using the Euclidean algorithm is $\mathcal{O}(B^3)$ [18]. The above algorithm requires $\frac{s(s+1)}{2}$ inversions in order to determine a degree s interpolating polynomial. Therefore, the amount of bit operations needed to update an L level degree a key tree using the polynomial interpolation scheme is $\mathcal{O}(a^2LB^3)$.

2.7 Chapter Summary

In order to address the problem of managing keys for securing multicasts, we proposed a framework that is suitable for dynamic group environments. Advanced protocol operations that update the keys during member joins, member departures, and the transferal of access rights were built using basic protocol operations which we call protocol primitives.

We described several desirable features for a multicast key management scheme, and which our scheme satisfied. In particular, our architecture provides a method for renewing session keys and key encrypting keys needed to control access to content. By using either the basic protocol operations, or more advanced protocol operations, the session key or key encrypting keys can be refreshed when a key's lifetime expires due to age or changes in membership. It is also evident that if users were to collude, they would not be able to figure out keys that they did not have. Users may survive accidents or move across terminals by sending a request for reinsertion to the server, upon which the server performs the member reinsertion protocol operation. We also provided a description of a protocol operation that would allow users to transfer their access rights to other parties. The server can revoke access to an individual by using the member departure operation to remove the member from the key hierarchy. Finally, our protocol uses a tree-structured key hierarchy in order to achieve desirable communication requirements during changes in the group membership.

A novel feature of this scheme is that it uses polynomial interpolation in conjunction with a broadcast seed to handle member departure operations. We studied the communication associated with performing member join and member departure operations. It was observed that higher tree degrees are best for member join operations, whereas a tree degree of 3 or 4 was best for the member departure operation. When equally weighting the join and depart operations, a degree 4 tree stood out as optimal. The communication overhead of the polynomial interpolation scheme is reduced in comparison to a model conventional scheme. We provided a comparison between the communication overhead of our scheme and the overhead of an example conventional scheme that used ID messages to flag the users which parts of the rekeying message were intended for them. As group size and tree degree increased, the communication overhead for the conventional scheme increases and ultimately becomes more burdensome than sending the broadcast seed. For example, when the group size was $n = 100000$ and the tree degree was $a = 4$, the communication overhead in the conventional scheme was approximately 25 % more than the overhead associated with a broadcast seed of size $B_\mu = 64$ bits. Finally, if one uses the previous session key $K_s(t-1)$ as the seed $\mu(t)$, then no communication overhead is associated with our protocol during member departures.

We presented a study of the communication needed when using our architecture to perform member joins and member departures. These two operations are the most important operations that a multicast server will have to face when operating in dynamic environments. The communication requirements of the member join and member departure operations lead to conflicting tree design considerations. By explicitly computing these two quantities as functions of the degree of the tree and computing the communication overheads, we studied the tree selection criterion. From our computations, the communication during a member join is reduced when using a higher degree tree, while the optimal tree degree for a member departure is either $a = 3$ or $a = 4$. We considered the average of the communications for the two operations, which gave strong support to choosing $a = 4$ as the optimal tree degree. We presented a stochastic population model that allows one to study the mean behavior of our architecture for varying amounts of users. It is observed that for both the join and departure operation, the amount of communication needed to update the key tree rapidly increases as the tree approaches 10% population. Above 10% occupancy, the communication needed for both operation stabilizes. We also examined the computational requirements of the tree-based rekeying schemes using polynomial interpolation.

3

Group Key Agreement Techniques in Heterogeneous Networks

Prior to the delivery of data intended for a group of recipients, it is necessary to initially establish keying material used to secure the group application. In this chapter we investigate the initial key agreement problem for both homogeneous and heterogeneous networks, whereby the members of a group each make contributions to establishing secret information that may be used to form a group encryption key.

3.1 Introduction

As noted earlier, key distribution can be accomplished by using a centralized entity, who shares private keying material a priori with communication entities and is responsible for distributing session keys to users. However, in many cases prior keying material might not exist or it might not be desirable to use a centralized third party. This might occur in applications where group members do not explicitly trust a single entity, or no member has the resources to maintain, generate and distribute information by himself. In these cases, *contributory* approaches are needed, where the group members each make independent contributions to the formation of the group key.

The classic example of a contributory scheme is the Diffie-Hellman (DH) key establishment scheme [19], in which two parties exchange messages that allow them to securely agree upon a key that may be used to protect their two-party communication. Several researchers have studied the problem of establishing a *Diffie-Hellman like* conference key [20–25]. Typically, these conference key establishment schemes seek to minimize either the amount of rounds needed in establishing the group key, or the size of the message.

As we move towards future communication scenarios, users will interact with each other from a variety of different communication and computing platforms, across a variety of different network types. We should thus expect that many applications will involve a heterogeneous clientele consisting of group members with different computational capabilities, pricing plans, and bandwidth resources. For these applications, minimizing the total bandwidth or amount of rounds might not be an appropriate metric. Instead, one should aim to minimize a cost function that incorporates the different costs or resource constraints of each user. The key generation scheme must therefore decide whether it is feasible to generate a key and determine a procedure for generating the group key while minimizing the total cost subject to resource budget constraints.

In this chapter, we discuss several methods for efficiently establishing a Diffie-Hellman like conference key that address the heterogeneous requirements of the conference members. We start in Section 3.2 by reviewing the Diffie-Hellman protocol, and presenting several conference keying schemes that employ the Diffie-Hellman problem. In Section 3.3, we present the butterfly scheme, a conference keying scheme for a homogeneous group of users, which builds the group key using the approach of [20]. The butterfly scheme can be generalized and we show that an underlying tree, which we call the *conference tree*, governs the process by which subgroup keys are formed en route to establishing the group key. By examining different shapes of conference trees, a family of tree-based group DH schemes can be formed. In Section 3.4, we consider the problem of designing a conference tree that can address cases where the users have different capabilities. We first examine the case when the users have different costs. In this case, the optimal conference tree can be constructed using the Huffman algorithm. We then examine the problem of choosing a conference tree when the users have the same cost, but are subject to varying budget constraints. We present necessary conditions for the existence of a conference tree when the users have budget constraints, and present an algorithm that minimizes the total cost given the budget constraints. Next, we consider the more general case where the users have different costs as well as different budgets. A computationally efficient near-optimal algorithm is presented that determines a conference tree whose total cost is very close to the optimal performance achieved by conference trees determined using either full-search or integer programming techniques. In Section 3.5, we present the results of simulations comparing the cost of forming a group key using tree-based schemes and several existent schemes. We also present simulations comparing the likelihood that a group key can be formed given that the users' budgets are drawn according to different distributions. From these simulations we conclude that the tree formulation for establishing a group key allows for great flexibility, and can efficiently establish group keys in resource-limited scenarios. Finally, in Section 3.6, we study the effects that the quantization and clipping of user costs have upon the total cost, and then investigate the effect that untrusty users can have upon the total cost of forming the group key using the Huffman-based conference tree. By suitably choosing the appropriate threshold level in the clipping operator, the effects of greedy user behavior are ameliorated.

3.2   Group DH Overview

In the basic DH scheme, the operations take place in an Abelian group G, typically chosen to be $\mathbf{Z}_p$ (the integers mod a prime p), or the points on an elliptic curve under appropriate laws of addition [26]. For consistency of notation, we shall develop our results for the group $\mathbf{Z}_p$. A group element g is chosen such that g generates a suitably large subgroup of G (preferably the whole group). Both party A and party B choose a private secret $\alpha_j \in \mathbf{Z}_p^*$ where $j \in \{A, B\}$ and $\mathbf{Z}_p^*$ denotes the non-zero elements of $\mathbf{Z}_p$. They each calculate $y_j = g^{\alpha_j}$ and exchange y_j with each other. Party A then calculates the key via $K = (g^{\alpha_B})^{\alpha_A} = g^{\alpha_B \alpha_A}$ and similarly for party B.

The problem of establishing a *Diffie-Hellman like* conference key has been investigated by several others [20–22]. One of the first *Diffie-Hellman like* conference key establishment schemes was proposed by Ingemarsson et al [20]. In the Ingemarsson (ING) scheme, the group members are arranged in a logical ring (e.g. $A \rightarrow B \rightarrow C \rightarrow A$). In a given round, every participant receives a message from its left-hand neighbor, raises that to their exponent, and passes it to their right-hand neighbor. For example, in the first round of a three person group exchange, we have $A \rightarrow B : g^{\alpha_A}$, $B \rightarrow C : g^{\alpha_B}$ and $C \rightarrow A : g^{\alpha_C}$. Then, in the second round $A \rightarrow B : (g^{\alpha_C})^{\alpha_A}$, $B \rightarrow C : (g^{\alpha_A})^{\alpha_B}$, and $C \rightarrow A : (g^{\alpha_B})^{\alpha_C}$. Finally, the shared key is $g^{\alpha_A \alpha_B \alpha_C}$, which they each can calculate by raising the final received message to their private exponent. For n users this scheme requires $n - 1$ rounds.

Another notable scheme is the Burmester-Desmedt conference key establishment scheme [21]. This scheme consists of three rounds. During the first round, each user u_j generates a random exponent α_j and broadcasts $z_j = g^{\alpha_j}$. The second round consists of each user u_j receiving z_j and broadcasts the quantity $x_j = (z_{j+1} z_{j-1}^{-1})^{\alpha_j}$. In the final round, each user u_j calculates the shared key

$$K = z_{j-1}^{n\alpha_j} x_j^{n-1} x_{j+1}^{n-2} \cdots x_{j-2}.$$

It can be shown that the shared key is actually the quantity

$$K = g^{\alpha_1 \alpha_2 + \alpha_2 \alpha_3 + \cdots \alpha_n \alpha_1}.$$

In [22], the GDH.1, GDH.2 and GDH.3 protocols are described that extend the two-party DH scheme to the n-party case. The distinguishing characteristic of the GDH.1/2 protocols is that they consist of two stages: an upflow and a downflow stage. For example, in the upflow stage of protocol GDH.1 user u_j receives a message of the form

$$\{g^{\alpha_1}, g^{\alpha_1 \alpha_2}, \cdots, g^{\alpha_1 \cdots \alpha_{j-1}}\}$$

and computes $g^{\alpha_1 \alpha_2 \cdots \alpha_j}$ by taking the last element of the received message and raising it to the α_j power. User u_j then sends to user u_{j+1} the message

$$\{g^{\alpha_1}, g^{\alpha_1 \alpha_2}, \cdots, g^{\alpha_1 \cdots \alpha_{j-1}}, g^{\alpha_1 \cdots \alpha_j}\}.$$

During the downflow stage, user u_n takes the output of the upflow stage, treats $g^{\alpha_1 \cdots \alpha_n}$ as the key, calculates g^{α_n} and raises the first $n-2$ elements of the output of the upflow stage to the α_n power. Then user u_n sends user u_{n-1} a message of the form $\{g^{\alpha_n}, g^{\alpha_1 \alpha_n}, \cdots, g^{\alpha_1 \cdots \alpha_{n-2} \alpha_n}\}$. User u_j performs likewise, calculating the key $g^{\alpha_1 \cdots \alpha_n}$ using the last term of the received message, and forwards to u_{j-1} a message formed by taking the first $j-1$ terms of the received message and raising them to the α_jth power. The GDH.3 scheme is a *centralized* Diffie-Hellman style scheme that differs from GDH.1/2 in that one user gathers contributions from all users, performs the majority of the computation for the group, and sends messages to each user that can be used to calculate the group secret. The *centralized* nature of the GDH.3 scheme is a drawback in environments where there is no single entity with significantly more computational capabilities than the others users. An extension to the GDH schemes that incorporates user authentication was presented in [24].

Several measures have been proposed to gauge a conference key protocol's complexity [22, 23]. The amount of messages sent and received, as well as the amount of bandwidth consumed are important measures of a protocol's efficiency. Another important measure that arises is the amount of rounds that a protocol takes in order to establish a group secret. A protocol that takes more rounds to establish a shared key is less favorable in environments where time and synchronization are precious resources. In [23], the communication complexity involved in establishing a group key is studied. In this work, lower bounds for the total number of messages exchanged, as well as the amount of rounds needed to establish the group key, were determined. They further present a key establishment scheme based upon a hypercube structure where the amount of rounds needed to establish the key is logarithmic in the group size.

A similar technique was proposed in [27, 28], in which the problem of group key establishment was examined in terms of signal flow graphs. The basic approach, called the *butterfly* scheme, had communication flow that was reminiscent of the butterfly diagrams of FFT calculations. The butterfly scheme used the ING scheme as the basic building block, and provided a broad family of approaches in which the amount of rounds needed to establish the group key is logarithmic in the group size. We will examine the butterfly scheme in the following section.

3.3   Conference Trees and the Butterfly Scheme

The general butterfly scheme is built using the ING scheme. However, since the two-party DH protocol is a special case of the ING scheme, we shall use the two-party DH protocol to introduce the basic ideas involved and then extend to using more general ING schemes. We refer to butterfly schemes built using two-party DH as radix-2 butterfly schemes. The terms *radix* and *butterfly* are borrowed from the signal processing community, and their usage is motivated by the resemblance between the communication flow of our butterfly scheme, and the butterfly signal flow diagrams associated with FFT computations [29]. In our work, the usage of radix refers to the size of the initial subgroups used in the butterfly scheme.

In order to explain the basic idea behind the radix-2 butterfly scheme, suppose that the number of users n is a power of 2. The users are paired up with each other to form two-person subgroups, and a key is established for each of these two-person subgroups using the conventional DH protocol. These subgroups are paired up with each other to form larger 4 member subgroups, and the two-party DH protocol is used to establish a group key for the 4 member subgroups. We may successively group subgroups to form larger subgroups and use two-party DH to ultimately achieve a shared group key.

A formal description of the butterfly scheme for $n = 2^r$ members is as follows. Initially, suppose each user u_j has a random secret integer $\alpha_j \in \mathbf{Z}_p^*$. The n users are broken into pairs of users $u_j^1 = \{u_{2j-1}, u_{2j}\}$. Here we have used the superscript in the notation to denote which round of pairings we are dealing with, while the subscript references the pair. We also refer to the initial secrets that each user possesses as $x_j^0 = \alpha_j$. In the first round, the members of a pair exchange their calculated $g^{x_j^0}$. For example, u_1 sends $g^{x_1^0}$ to u_2, and u_2 sends $g^{x_2^0}$ to u_1. Then, the users u_{2j-1} and u_{2j} each calculate $x_j^1 = g^{x_1^0 x_1^0} = g^{\alpha_{2j-1}\alpha_{2j}} \pmod{p}$. Since $x_j^1 \in \mathbf{Z}_p^*$, and both members of a pair have established a conventional DH key, we may now group the pairs u_j^1 into a second level of pairs, e.g. $u_1^2 = \{u_1^1, u_2^1\}$, and more generally $u_j^2 = \{u_{2j-1}^1, u_{2j}^1\}$ so that the second level of pairings consists of 4 users in a pair. Each user from u_{2j-1}^1 has an associated member of u_{2j}^1 to whom they send $g^{x_{2j-1}^1}$ and similarly receive $g^{x_{2j}^1}$ from. Every member in u_j^2 can calculate $x_j^2 = g^{x_{2j-1}^1 x_{2j}^1} \pmod{p}$. A third pairing, consisting of 8 users may be formed and a similar procedure carried out if needed. In general, $u_j^k = \{u_{2j-1}^{k-1}, u_{2j}^{k-1}\}$ and $x_j^k = g^{x_{2j-1}^{k-1} x_{2j}^{k-1}} \pmod{p}$. Ultimately, the procedure continues until there are only two intermediate values that can be combined to get the group secret.

A trellis diagram depicting the communication flows between users is depicted in Figure 3.1 (a). It is not necessary that each user perform a communication during each round. In fact, such an operation might use more power since many users are transmitting identical information. In

---

for *Stage $k = 1 : r$* **do**

  Form subgroups $u_j^k = \{u_{p_k(j-1)+1}^{k-1}, u_{p_k(j-1)+2}^{k-1}, \cdots, u_{p_k j}^{k-1}\}$;

  Establish a secret x_j^k for subgroup u_j^k using x_j^{k-1} as secrets in a p_k-member ING scheme ;

end

Algorithm 2: Algorithm for calculating the group key using ING scheme when the group size is factored as $n = p_1 p_2 \cdots p_r$.

networks, such as wireless networks, where broadcasting is available, alternative trellis diagrams can be constructed where one user broadcasts an intermediate message to multiple users. An example of such a trellis is depicted in Figure 3.1 (b). An alternative way to view the butterfly scheme is provided in Figure 3.1 (c), which depicts the tree associated with the butterfly scheme. This tree, which we refer to as the *conference tree*, describes the successive subgroups and subgroup keys that are formed en route to establishing the key for the entire group. For example, there is a node on the conference tree that is the grandparent of $\{u_1, u_2, u_3, u_4\}$ and hence there is a subgroup key that can allow $\{u_1, u_2, u_3, u_4\}$ to communicate securely amongst themselves if so desired.

When n is not a power of 2, a group key still can be established easily. In this case, we form a subgroup with an amount of users equal to the largest power of 2 less than or equal to n. The remaining users are further broken down in a similar fashion, resulting in a new set of remaining users that can be further broken down. For example, a group of 7 users will be broken down into subgroups of 4, 2, and 1 members. Subgroup and group keys are formed in a fashion similar to the case when n is a power of 2. The trellis and conference tree for $n = 7$ users is depicted in Figure 3.2. The number of rounds needed to complete the radix-2 butterfly scheme is $\lceil \log_2 n \rceil$.

We now extend the approach used above to employ the more general ING scheme as the basic building block. Since the resulting schemes are not built using a two-party protocol, they are termed non-radix-2 butterfly schemes. Suppose that $n = p_1 p_2 \cdots p_r$ is the number of users, and the p_j are not necessarily prime. The general ING butterfly scheme starts by breaking the group into subgroups of size p_1 and uses the ING scheme to establish a shared key for each of the $n_2 = p_2 \cdots p_r$ subgroups. The n_2 subgroups are further broken down into subgroups consisting of p_2 subgroups, and the ING protocol is used to establish subgroup keys for these larger subgroups. The process continues until a key is established for the entire group. The procedure for this scheme is presented in Algorithm 2, where $u_j^0 = u_j$ and the initial user secrets are $x_j^0 = \alpha_j$. An example is depicted for the case of $n = 9$ users in Figure 3.3. The total amount of rounds is $2 \log_3 9 = 4$, and the amount of messages is 36. The direct use of the ING scheme for 9 users requires 8 rounds and 72 messages. The divide and conquer strategy in the butterfly approach improves the efficiency of the ING scheme. Additionally,

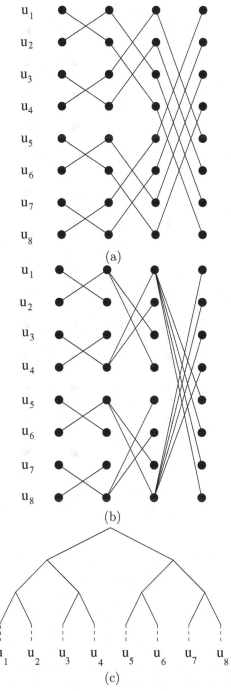

FIGURE 3.1. The radix-2 butterfly scheme for establishing a group key for 8 users.
(a) Without broadcasts, (b) Using broadcasts, and (c) the associated conference
tree.

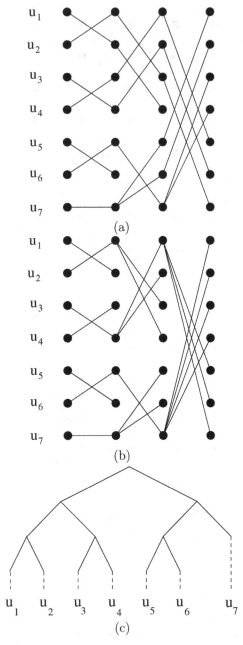

FIGURE 3.2. The radix-2 butterfly scheme for establishing a group key for 7 users. (a) Without broadcasts, (b) Using broadcasts, and (c) the associated conference tree.

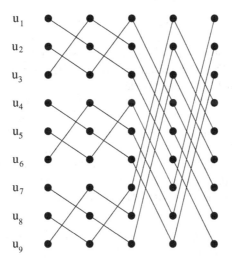

FIGURE 3.3. The trellis for $n = 9$ users using two levels of 3-party ING scheme.

the logarithmic amount of rounds needed by the butterfly scheme to establish the group key is an improvement over the linear amount of rounds required by the GDH schemes of [22]. We further note that the hypercube approach of [23] also requires a logarithmic amount of rounds to establish the conference key. However, the hypercube approach does not address the issue of using a general subgroup size as the building block for designing a scalable conference key establishment scheme. By using the ING scheme as the basic module in the butterfly scheme, we include the hypercube approach as a special case, and have generalized their approach. Further, the butterfly scheme described above allows for the use of multicast channels to improve communication efficiency.

It is not necessary to use a factorization of n in designing the non-radix-2 butterfly scheme. In fact, for prime n, this factorization would necessitate using an n-party ING scheme, and require a large amount of rounds in forming the group key. Rather, what is required is that the degrees p_j of the ING schemes used satisfy $\prod p_j \geq n$. In this case, some positions are left unused. For example, when $n = 8$ and $p_1 = p_2 = 3$ one position of a 3-party ING scheme is empty, in which case that computation simply uses the 2-party DH scheme instead

The total number of rounds needed in the ING butterfly scheme for $n = p_1 p_2 \cdots p_r$ users is

$$TR = \sum_{j=1}^{r} (p_j - 1) = \left(\sum_{j=1}^{r} p_j \right) - r. \tag{3.1}$$

When choosing a factorization to represent n, the more factored representation leads to a smaller number of rounds TR. We now show that using a binary conference tree produces the group key in the fewest amount of rounds. To do this, we show that if one uses a p_j ING scheme for round j of the group key establishment, then the use of several two-party DH schemes in place of the p_j ING scheme either produces the same amount of rounds or fewer in establishing the group key.

If we require that all of the computation on one level of a conference tree is completed prior to the formation of the keys in the next level up the conference tree, then using the two-party DH scheme as the building block leads to trees with the least amount of rounds needed to establish the group key. The proof for this claim is provided in the following lemma. Since using two-party DH leads to binary trees that require the least amount of time rounds, we shall restrict our attention to binary trees for the remainder of the chapter.

Lemma 3. *Let n be the amount of users, and suppose that we wish to establish a conference tree where level j uses a p_j ING scheme as the basis, then a binary tree (where $p_j = 2$) produces an optimal conference tree.*

Proof. Suppose that you have an optimal set of numbers $\{p_1, \cdots, p_r\}$ that are used to construct the conference tree for n users. Then the number $N = \prod_{j=1}^r p_j \geq n$, and the total rounds $TR = \left(\sum_{j=1}^r p_j\right) - r$ is minimal.

We will show that if there is a $p_j \neq 2$ then we may replace p_j by a sequence of numbers all of which have value 2. Suppose there is a j such that $p_j \neq 2$, then the p_j contributes $\Delta_j = p_j - 1$ to the total amount of rounds TR. Define $p'_j = \{2, 2, \cdots, 2\}$ which is a sequence of length $\lceil \log_2 p_j \rceil$. If we use this set of numbers in place of p_j, we instead contribute $\Delta'_j = \lceil \log_2 p_j \rceil$ to the total cost. It is clear that using p'_j in place of p_j produces an $N' \geq n$. However, the incremental cost $\Delta'_j = \lceil \log_2 p_j \rceil$ is less than or equal to Δ_j (in fact, if $p_j = 3$ then equality holds, else it is strictly less). Thus, if $p_j > 3$ then replacing p_j by p'_j produces a set of numbers with lesser amount of total rounds TR, which contradicts optimality. On the otherhand, if $p_j = 3$ then replacing p_j by p'_j will produce a set of numbers with an equal amount of total rounds TR, and hence we may choose to use p'_j instead of p_j in the construction of the optimal tree. By applying this argument to all $p_j \neq 2$ we conclude that a binary tree must produce an optimal tree.   □

It should be pointed out, however, that the argument used above does not produce the optimal tree, but rather only implies that the optimal tree is binary. For example, consider $n = 27$. The total amount of rounds using three levels of 3-party ING is $TR = 6$. If we use the above technique, we replace each 3 by $2 \cdot 2$, and get a conference tree with 2^6 terminal nodes and total cost of 6. However, the optimal tree in this case is the binary tree of depth $\lceil \log_2 n \rceil$, with total rounds $TR = 5$.

In the butterfly schemes described above, the conference trees were almost balanced and full. For example, the conference tree for $n = 8$ users involves 3 levels of internal nodes, and all 8 users are placed at the same depth in the tree. For more arbitrary amounts of users, the users are all roughly placed at the same depth. More general depth assignments and conference tree structures may be given to the users. In the next section, we shall exploit the extra freedom provided by more general binary conference trees by placing users at different depths in order to reduce the total group cost needed to form the group key.

3.4   Computational Considerations

In many application environments users will have varying amounts of computational resources available. Low-power devices, such as wireless appliances, cannot be expected to expend the same amount of computational effort as high-power devices, such as personal computers, when establishing a group secret. It is therefore important to study the problem of efficiently establishing a conference key while considering the varying user costs.

To accomplish the efficient establishment of a conference key in a heterogeneous environment, we introduce a new entity, called the Conference Keying Assistant (CKA). The CKA is responsible for collecting the users' costs or budgets, determining the appropriate conference keying tree, and conveying the conference tree to the conference members if it is feasible to establish the group key. The CKA is not responsible for performing any computation beyond the calculation of the appropriate conference tree, and therefore only needs to be a *semi-trusted* entity who will accurately convey the conference tree to the conference members. We note that the CKA may be a member of the conference, in which case his duties as CKA are in addition to his role as a group member.

In this section, we present methods that the CKA can employ to design the conference tree that is used by the group members to establish the group secret. In particular, we study two problems: minimizing the total cost in establishing a group key, and the feasibility of establishing the group key in the presence of budget constraints. We present algorithms to efficiently determine the conference keys for each of these problems separately, and then together.

3.4.1   Minimizing Total Cost

First, assume that we have n users, and that each user u_j has a cost $w_j \geq 0$ associated with performing one two-party Diffie-Hellman protocol. For example, this cost might be related to the amount of battery power consumed. Suppose we place the n users on a conference tree with n terminal nodes in such a manner that each user u_j has a length l_j from his terminal node to the root of the conference tree. Our goal is to minimize the total cost $C = \sum w_j l_j$ of this tree.

We first address the question of what is the minimum amount of total computation necessary for establishing the group key for n users. This problem can be addressed using coding theory. If we define p_j as $p_j = w_j/(\sum_k w_k)$, then $\sum_j p_j l_j$ is just a scaling of $\sum_j w_j l_j$ by $W = \sum_k w_k$. Let us define X to be a random variable with a probability mass function given by p_j, then minimizing $\sum_j p_j l_j$ is equivalent to finding a code for X with lengths l_j that minimizes the average code length. We thus infer the following lower bound on the total cost for establishing a group key, which follows from the lower bound for expected codelength of an instantaneous binary code for X [30]:

Lemma 4. *Suppose that n users wish to establish a group secret and each user u_j has a cost w_j associated with performing one two-party Diffie-Hellman protocol. Then the total cost C of establishing the group secret satisfies $-W \sum_j p_j \log_2 p_j \leq C$ where $p_j = w_j/W$.*

The observation that efficiently establishing a group key is related to coding allows the CKA to use procedures from coding theory to determine desirable conference trees. In particular, Huffman coding [31–34] is computationally efficient and yields the optimal conference tree that minimizes the total weighted cost. That is, if C^* is the cost of forming the group key using the Huffman tree, then the cost C' of using a different conference tree assignment will satisfy $C' \geq C^*$. Since Huffman coding produces an optimal code, we know that the expected cost $\sum_j w_j l_j^*$ satisfies the following bound

$$WH(p) \leq \sum_j w_j l_j^* < W\left(H(p) + 1\right), \qquad (3.2)$$

where $H(p)$ is the entropy of the distribution p. Thus, the Huffman construction of the conference key tree has a total cost that is within W of the lower bound.

The following example demonstrates the advantage of using the Huffman algorithm for forming the conference tree when compared to using the full balanced tree of the radix-2 butterfly scheme.

Example 1. *Consider a group of 8 users with costs $w_1 = 28$, $w_2 = 25$, $w_3 = 20$, $w_4 = 16$, $w_5 = 15$, $w_6 = 8$, $w_7 = 7$, and $w_8 = 5$. The Huffman algorithm yields the tree depicted in Figure 3.4. The corresponding length vector is $l^* = (2, 2, 3, 3, 3, 4, 5, 5)$, and the total cost is 351. The total cost for a full balanced tree is 372.*

We now quantify the improvement that is available when using the Huffman code compared to the cost of using an arbitrary conference tree. For an arbitrary conference tree, we suppose that the length assigned to user u_j is l_j. If we define a probability distribution q by $q_j = 2^{-l_j}$, then the expected

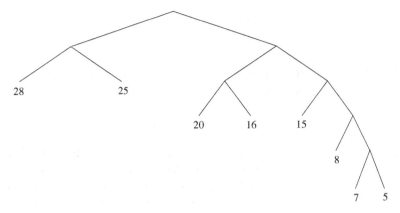

FIGURE 3.4. Huffman example

length under the probability p_j of the code with lengths l_j satisfies [30]

$$H(p) + D(p\|q) \leq \sum_{j=1}^{n} p_j l_j < H(p) + D(p\|q) + 1. \tag{3.3}$$

Here $D(p\|q)$ is the Kullback-Leibler divergence between the two probability distributions p and q. The cost for using this tree is $C = W \sum p_j l_j$. We can combine the bound of Equation (3.3) with the bound for the cost of the optimal code $C^* < W(H(p) + 1)$ to get $C - C^* > W(D(p\|q) - 1)$. When $D(p\|q) > 1$, this bound is an improvement over the trivial bound $C - C^* \geq 0$.

3.4.2 Budget Constraints

In many cases, the parties wishing to establish a conference key might have a limited budget to spend. The optimal conference tree assignment that results from Huffman coding might assign more computation to some users than they are capable of performing, while assigning less computation to other users than they are capable of performing. In these cases, rather than minimize the total cost, one should ensure that one can first establish the group key, and then reduce the total amount of computation as a secondary issue.

Suppose that user u_j publishes a budget b_j that describes the amount of two-party Diffie-Hellman key establishment protocols he is willing to participate in when establishing the group key. Without loss of generality, we assume that the users' budgets b_j satisfy $b_j \leq b_k$ for $j < k$. We define the budget vector as $b = (b_1, b_2, \cdots, b_n)$. The length vector $l = (l_1, l_2, \cdots, l_n)$ describes the lengths from each user's node to the root of the conference tree. The necessary conditions on the budget vector b for the existence of a conference key tree with lengths $l_j \leq b_j$ is provided by the Kraft Inequality [30]:

Lemma 5. *Suppose that the budget vector $b = (b_1, b_2, \cdots, b_n)$. Then a conference key tree with lengths l_j exists that satisfies the budget constraint $l_j \le b_j$ for all j if $\sum_{j=1}^{n} 2^{-b_j} \le 1$.*

A budget vector that satisfies the Kraft Inequality is said to be *feasible*. When a budget assignment does not satisfy the Kraft Inequality and we choose to drop a single member to generate a feasible budget vector for the remaining users, then the best strategy is to drop the member with the lowest b_1.

Using the budget vector as the length vector does not always lead to a full conference tree in which every node has two children. In order to get a full tree, we must trim the budget vector to produce a length vector l that achieves the Kraft Equality. The length vector is formed by reducing elements of the budget vector by amounts that do not violate the Kraft Inequality. The following lemma provides a useful approach to trimming the length vector assignment while still satisfying the Kraft Inequality.

Lemma 6. *Suppose $b = (b_1, b_2, \cdots, b_n)$ with $b_j \le b_k$ for $j < k$ satisfies the strict Kraft Inequality, $\sum 2^{-b_j} < 1$, then the modified budget vector c defined by $c = (b_1, b_2, \cdots, b_{n-1}, b_n - 1)$ satisfies the Kraft Inequality $\sum 2^{-c_j} \le 1$.*

Proof. Observe that 2^{b_n} is the least common denominator of the set 2^{-b_j}. Thus $\sum 2^{-b_j}$ can be expressed as

$$\sum 2^{-b_j} = \frac{x_1 + x_2 + \cdots + x_n}{2^{b_n}} < 1 \qquad (3.4)$$

where $x_j = 2^{b_j - b_n}$. In particular, $x_1 + x_2 + \cdots + x_n < 2^{b_n}$, and as a consequence $x_1 + x_2 + \cdots + (x_n + 1) \le 2^{b_n}$. However, $(x_n + 1)/2^{b_n} = 1/2^{b_n - 1}$, and so the sequence $(b_1, b_2, \cdots, b_n - 1)$ satisfies the Kraft Inequality. $\square$

A consequence of this lemma is that if we subtract 1 from one of the b_j, then choosing the largest b_j least affects $\sum_j 2^{-b_j}$. Using this idea, Algorithm 3 starts with an admissible budget vector b, initializes the length vector $l = b$, and produces a length assignment $l = (l_1, l_2, \cdots, l_n)$ satisfying $l_j \le b_j$ such that $\sum_j 2^{-l_j} = 1$ and $\sum_j l_j$ is minimized over all length vectors c satisfying $\sum_j 2^{-c_j} \le 1$. The optimality of this algorithm is discussed in Lemma 7.

As an example of the algorithm, suppose $n = 8$ and that the initial budget is $b = (1, 3, 3, 4, 5, 5, 6, 8)$. This budget vector is feasible and performing the algorithm gives the final assignment $l = (1, 3, 3, 4, 4, 4, 5, 5)$.

Lemma 7. *Algorithm 3 produces an optimal length assignment vector l to the problem*

$$\left\{ \min_{l} \sum_{j} l_j \; : \; 1 \le l_j \le b_j, \sum_{j} 2^{-l_j} \le 1, l_j \in \mathbf{Z}^+ \right\}. \qquad (3.5)$$

Data: A length vector l satisfying $\sum 2^{-l_j} \leq 1$.
while $\sum 2^{-l_j} < 1$ **do**
  $j = \arg\max\{l_k\}$;
  $l_j = l_j - 1$;
end

Algorithm 3: Algorithm for calculating the optimal length vector l.

Proof. We will aim to show that there is an optimal solution in which one decreases the largest value of the budget vector by one. Let l^* be an optimal solution to the problem. Then by the previous lemma $\sum 2^{-l_j^*} = 1$. Consider a sequence of steps that take the budget vector b to the optimal length vector l^* by decreasing one element by 1 during each step. We denote by J_* the sequence of indices involved in going from b to l^*, where $J_*(k)$ refers to the index of the budget vector that is decreased during kth step. Let j_0 be the index of the largest element of b, we claim there is an optimal solution l' with a corresponding $J_*(1) = j_0$. If $J_*(1) = j_0$ then we are done. However, if $J_*(1) \neq j_0$ then there are two cases. The first case is that there is another element of J_* that has value j_0, in which case we may switch that element with $J_*(1)$ to produce a new sequence of steps that does not alter the value of $\sum 2^{-l_j^*}$ and maintains the optimality of $\sum_j l_j$. The second case is that $j_0 \notin J_*$. If there are any other elements of b with the same value as b_{j_0}, then indices of these may be used in place of j_0, and considered in the preceding argument. However, if there are no b_j's with the same value as b_{j_0} then we seek a contradiction as to the optimality of l^*. Choose an arbitrary element of J_*. This element, which we denote by $J_*(k)$, by assumption has the property that it $b_{J_*(k)} < b_{j_0}$. Define $J_*^- = \{J_*(1), \cdots, J_*(k-1), J_*(k+1), \cdots\}$, which corresponds to the sequence of steps involved in J_* excluding the kth step. Define $J_\sharp = j_0 \| J_{*-}$, which describes a new sequence of steps that starts with j_0 and then the steps of J_{*-}. Then $J_\sharp$ leads to a length vector $l^\sharp = l^* + e_{J_*(k)} - e_{j_0}$, where e_j is the vector of all zeros except in the jth index which has value 1. This length vector has the property that $\sum 2^{-l_j^\sharp} < \sum 2^{-l_j^*}$ since $2^{-l_{j_0}^*} < 2^{-l_{J_*(k)}^*}$. Hence $\sum 2^{-l_j^\sharp} < 1$. However, by the preceding lemma, this means that $l^\sharp$ can be used to produce a better length vector, which contradicts the optimality of l^*.

Hence, the optimal solution may as well have the first step reduce the largest element of the budget vector. Now the problem reduces to finding an optimal solution to the new budget $b' = b - e_{j_0}$. By induction on the number of steps, we therefore conclude that choosing the largest element during each step yields an optimal solution, and hence the greediness of Algorithm 3 is optimal. $\qquad\square$

3.4.3 Combined Budget and Cost Optimization

We have studied the problem of minimizing the total cost of establishing a group key using a tree structure, and whether a group key can be established in a budget-limited scenario. We now address the more realistic scenario where users have different costs as well as budget constraints. We are therefore interested in the problem of minimizing the total cost of the length assignments l_j for the weights w_j given the budget constraint $l_j \leq b_j$. This problem is formally stated as:

$$\text{Minimize} \qquad \sum_{j=1}^{n} w_j l_j$$

$$\text{subject to} \qquad 1 \leq l_j \leq b_j, \quad \sum_{j=1}^{n} 2^{-l_j} = 1, \quad l_j \in \mathbf{Z}^+$$

where $\mathbf{Z}^+$ denotes the non-negative integers. Once a length vector has been determined, it can be sorted in ascending order to describe a conference tree.

This problem is more difficult than either the minimum cost problem or the budget-constrained problem. If the budget vector is constant, i.e. $b_j = b$ for every j, then the methods of length-constrained source codes may be applied [35–41]. One efficient algorithm for finding the optimal Huffman code under the maximum codeword length constraint is presented in [35], which is based on the algorithm of [37]. A near optimal solution can be found using Lagrange relaxation, and an efficient implementation is described in [38]. However, in the more general case where the budgets vary from user to user, it is difficult to find the optimal solution since the ordering $w_j \leq w_k$ does not imply $l_j \geq l_k$.

Two suboptimal approaches that employ a greedy strategy were developed to tackle the general problem where the budgets vary from user to user. The first algorithm, described in Algorithm 4, is a variant of Algorithm 3, which starts with a length assignment $l = b$ and chooses to decrease the element l_j of the length vector that most reduces the total cost $\sum w_k l_k$ at that step while maintaining the Kraft Inequality. This greedy algorithm is not optimal, as can be seen by the example $b = (2, 2, 3, 3)$ with costs $w = (10, 7, 6, 6)$. In this example, the algorithm produces the length vector $l = (1, 2, 3, 3)$ (which has a total cost of 60), whereas the optimal length vector is $l^* = (2, 2, 2, 2)$ (which has a total cost of 58).

Algorithm 4 is a naive greedy algorithm. By slightly altering this algorithm, another greedy algorithm may be developed with better performance. Instead of decreasing the element that best decreases the total cost, Algorithm 5 chooses to decrease the element with the largest value $w_j 2^{l_j}$. This corresponds to choosing the element that would have the largest change in the cost function per change in the Kraft Inequality. A similar strategy is often used in designing incremental resource allocation schemes

---

Data: A budget vector b.
if $\sum 2^{-b_j} > 1$ **then**
   | No solution. ;
end
$l = b$;
while $\sum 2^{-l_j} < 1$ **do**
     Let $\delta = 1 - \sum 2^{-l_j}$;
     $K = -\lceil \log_2 \delta \rceil$;
     $J = \{j : l_j \geq K\}$;
     Let $j_0 = \arg\max_{j \in J} w_j$;
     $l_{j_0} = l_{j_0} - 1$;
end

Algorithm 4: Algorithm for calculating the length vector l, given budget b and costs w_j.

Data: A budget vector b.
if $\sum 2^{-b_j} > 1$ **then**
   | No solution. ;
end
$l = b$;
while $\sum 2^{-l_j} < 1$ **do**
     Let $\delta = 1 - \sum 2^{-l_j}$;
     $K = -\lceil \log_2 \delta \rceil$;
     $J = \{j : l_j \geq K\}$;
     Let $j_0 = \arg\max_{j \in J} w_j 2^{l_j}$;
     $l_{j_0} = l_{j_0} - 1$;
end

Algorithm 5: Improved algorithm for calculating the length vector l, given budget b and costs w_j.

in operations research [42]. Algorithm 5 is also suboptimal, but exhibits better performance than Algorithm 4 with a negligible increase in the amount of computation needed. The optimal solution to the combined budget and cost optimization problem can be obtained by performing either full-search, or using the methods of integer programming [43–45]. One useful approach is to apply the branch and bound method to the problem [43, 46–48].

We now compare the near-optimal results of Algorithm 5 with the optimal solution. We performed a simulation where each user's budget b_j was chosen uniformly from $[1, 3n]$, and weights w_j were chosen uniformly from $[1, 100]$. The optimal solution, l^*, was compared with the approximate solution, $\hat{l}$, from Algorithm 5 via the relative difference

$$\rho = \frac{C(\hat{l}) - C(l^*)}{C(l^*)}. \tag{3.6}$$

TABLE 3.1. Comparison between the optimal solution and the approximate solution of Algorithm 5 for different group sizes n.

Group size n	$\bar{\rho}$
5	0.0037
6	0.0046
7	0.0027
8	0.0020
9	0.0025
10	0.0020
11	0.0016

This quantity was calculated and averaged over 100 realizations to produce the mean relative difference $\bar{\rho}$, for the group sizes $n = 5, 6, 7, 8, 9, 10$, and 11. The results are presented in Table 3.1, which indicates that Algorithm 5 produces the group key with cost that is within 0.5% of the optimal cost. Due to the computational complexity required to find the optimal solution for 100 realizations, we only present results through $n = 11$.

Since determining the optimal solution is very computationally intensive for large group sizes, it is unreasonable for the CKA to find the optimal conference tree when users have both budget constraints and varying costs. Instead, Algorithm 5, although not optimal, has very competitive performance and its computational requirements are small compared to full-search or the branch and bound method, and is a reasonable candidate for the CKA to use in determining the conference tree.

3.5   Efficiency and Feasibility Evaluation

We now compare our tree-based conference key establishment schemes with other schemes in the literature. We assume that no broadcast channels are available, and that if one user desires to communicate amongst many, he must establish many separate connections. There are two evaluations that we present: first, we consider the total cost needed to establish a group key when the users have different costs; second, we examine the feasibility of establishing a conference key when group members have different budget constraints.

3.5.1   Comparison of Total Cost

We simulated a scenario in which there were three classes of users. The first class corresponds to users with a large amount of computational power (and hence lower user cost), the second corresponds to a medium level of computational power, and the last class represents users with low-powered devices or a high cost. In order to represent this distinction, the users were assumed to have weights drawn according to three different distributions. For every 10 users, 2 users have weights drawn according to the first distribution, 5 according to the second distribution, and 3 according to the third distribution. The first weight distribution was a discrete uniform distribution with integer values from $[1, 50]$, while the second was a discrete uniform distribution over $[501, 550]$, and the third was a discrete uniform distribution over $[951, 1000]$.

We compared the total cost for the Huffman scheme with the cost of the butterfly scheme, the ING scheme, the GDH.1/2 scheme, and the GDH.3 scheme. Since there are differences between the communication and computational procedures of the different schemes, we assume that the user costs are associated with the cost to perform the two modular exponentiations needed in a two-party DH scheme. This means, for example, that if a user has a cost of w to perform one round of two-party DH, then he has a cost of $3w/2$ to perform a 3-party ING scheme since there are 3 modular exponentiations involved.

We also assume that every user in a DH scheme performs the two modular exponentiations. For example, if the subgroup $\{u_1, u_2\}$ share a secret x and the subgroup $\{u_3, u_4\}$ share a secret y, and use DH to establish a shared key for the 4 members, then both u_1 and u_2 calculate g^x and g^{yx}. Similarly, both u_3 and u_4 calculate g^y and g^{xy}. In actuality, however, only one member from each subgroup must calculate and transmit the message g^x or g^y. The costs for the Huffman and butterfly schemes that we report do not reflect this possible savings, and are therefore overestimates of the actual costs.

The total cost required to establish the conference key was calculated for different group sizes and averaged over 500 realizations. The average costs are depicted in Figure 3.5. Examining Figure 3.5 we see that the ING and GDH.1/2 schemes have higher total cost than the Huffman, butterfly, and GDH.3 schemes. In this example, the Huffman scheme performs better than the butterfly scheme by an average of 6.7%. GDH.3 has the best performance in terms of total cost. However, GDH.3 is a *centralized* scheme and cannot be categorized as a completely *distributed* conference keying scheme since one user performs the majority of the computations for the group. In contrast, the Huffman scheme and the butterfly scheme are contributory and do not make any single user responsible for the majority of the computation (although they do allot more load to some users than others). In scenarios where it is appropriate to have one user or entity do nearly all of the work for the remaining users the use of centralized multicast key distribution schemes [8, 10, 49] will lead to more efficient distribution of keying information than conference keying schemes.

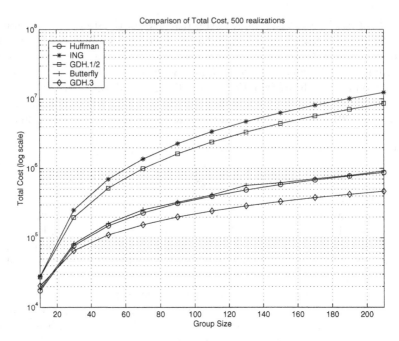

FIGURE 3.5. Cost comparison of establishing a conference key using the Huffman-based conference tree, the ING scheme, GDH.1/2, the butterfly scheme, and the GDH.3 scheme. The first four schemes are contributory protocols, while GDH.3 is a centralized protocol.

3.5.2 Feasibility Comparison

Another major concern is the feasibility of establishing a secure conference in the presence of budget constraints. When the users have different budgets, it might not be possible for different schemes to establish a conference key. We shall quantify the likelihood that a conference key can be established in a scenario where the users' budgets are drawn according to a distribution by introducing the PESKY(Probability of Establishing the Session KeY) measure.

Suppose that $\mathcal{B}$ denotes the set of all possible budget vectors for n users, and that μ is a probability distribution over $\mathcal{B}$ describing the likelihood of the users having a certain budget vector. Let a conference key scheme be denoted by F, and $F(\mathcal{B})$ the set of all budget vectors $\mathcal{B}$ which are feasible for F. Then formally, the PESKY measure is defined as:

$$PESKY(F, n) = \sum_{b \in F(\mathcal{B})} \mu(b). \tag{3.7}$$

For example, if we let F refer to a conference tree scheme built using Algorithm 3, Algorithm 4, or Algorithm 5, then a budget vector is feasible if it satisfies the Kraft Inequality, and therefore $F(B) = \{b : \sum_j 2^{-b_j} \leq 1\}$.

In general, it is difficult to find closed form expressions for PESKY, and Monte Carlo methods must be used to estimate PESKY.

We used PESKY to study the likelihood that different schemes could produce a group key when the user's budgets were drawn according to different distributions. We assumed that the budgets b_j correspond to the amount of two-party DH schemes that a user is willing to participate in, and that the two modular exponentiations are the most significant expense for the user. Therefore, each value of the budget allows for 2 modular exponentiations to be performed. As before, we assume that every user in a subgroup performs both of the modular exponentiations in a DH scheme. We compare the PESKY for Algorithms 3-5 with PESKY for both the GDH.1/2 and GDH.3 schemes for three different budget distributions. The PESKY for Algorithms 3-5 are conservative estimates of the actual probability of establishing the session key since it is not necessary that all members of a subgroup perform the first modular exponentiation of a DH scheme.

The first budget distribution is a discrete uniform distribution with integer values from $[5, 20]$. The distribution is presented in Figure 3.6(a), and the corresponding PESKY curves are presented in Figure 3.6(b). Since the GDH schemes require that one user performs an amount of modular exponentiations equal to the amount of users n, it is impossible for groups of more than 40 users to be formed via the GDH protocols with this distribution, as can be seen in Figure 3.6(b). The PESKY plots for this distribution demonstrate that it is more likely that a budget vector can satisfy the Kraft Inequality than the requirements of either the GDH.1/2 or GDH.3 schemes. In fact, it is not until the group sizes become larger than $n = 200$ that a significant decrease is observed in the likelihood of forming a group key using a conference tree.

For the second distribution, the elements of the budget vector are generated as $1 + NegBin(10, 0.25)$, where $NegBin(s, p)$ is the negative binomial distribution with probability mass function $q(b)$ given by:

$$q(b) = \binom{s + b - 1}{b} p^s (1 - p)^b \quad \text{for } b \in \{0, 1, \cdots\}. \tag{3.8}$$

The addition of 1 to $NegBin(10, 0.25)$ was to ensure that no users had a budget of 0. In Figure 3.7, we see that the tree-based schemes exhibit a 100% likelihood of successfully establishing the conference key for conferences with between 10 and 200 users. The PESKY values for GDH.3 begins to drop off at $n = 100$ users while the values for GDH.1/2 drop off at $n = 80$ users. Since a large amount of budget values are above 20, the PESKY curves do not drop off as quickly as they did for the uniform case.

In the third distribution, the budgets were drawn as $1 + NegBin(5, 0.3)$, as depicted in Figure 3.8 (a), and the corresponding PESKY measures are depicted in Figure 3.8 (b). This distribution describes a similar phenomenon to the uniform distribution above, but includes a heavier tail at higher budget values that could represent a diminishing class of more powerful

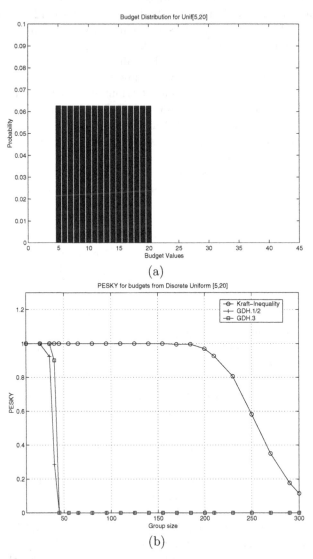

FIGURE 3.6. (a) Budget distribution discrete uniform with integer values from [5, 20] (b) Corresponding PESKY

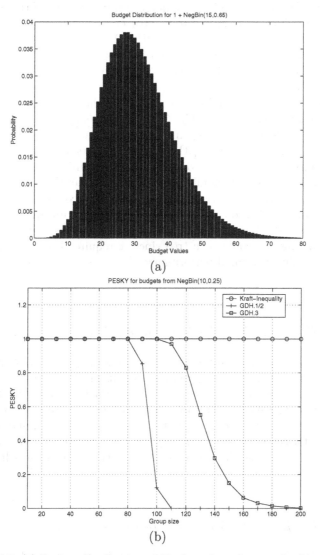

(a)

(b)

FIGURE 3.7. (a) Budget distribution, shifted version of a negative binomial distribution with parameters $s = 10$, and $p = 0.25$. (b) Corresponding PESKY

users. The fact that roughly 6% of this distribution corresponds to budget values below 5 has a significant effect upon the PESKY plots. In particular, we see that the PESKY all of the conference keying schemes drop off earlier than in Figure 3.6 (b). For example, when there are $n = 70$ users there is only an 80% chance of forming a conference key using one of these schemes with this distribution compared to a 100% chance with the distribution of Figure 3.6 (a). We also see that the GDH.1/2 schemes are very unlikely to successfully establish a group key, even for group sizes of $n = 20$, and that all of the GDH schemes are unable to establish a group key for groups of more then 60 users.

Therefore, in resource-limited scenarios, the choice of which conference keying scheme is very critical. The GDH.3 scheme, although cost-efficient, obtains this efficiency at the expense of requiring a single user have significantly more power and resources than the other users. In applications where the users have a more balanced distribution of resources, the GDH schemes have PESKY graphs that rapidly drop off and are therefore unlikely to successfully establish a group key. In these cases, the conservative estimates of PESKY for tree-based conference keying schemes indicate that they are more likely to establish a group key, and Algorithm 5 is a judicious choice for constructing the conference tree since it requires little computational effort and has near-optimal performance.

3.6   System Sensitivity to False Costs

In this section, we examine the effect that announcing costs different from the true user costs has upon the total cost of using the Huffman conference tree. There are two cases that we consider. First, we consider the issue that users announce costs that are approximations of the true costs. Next, we examine the case where some of the users are untrusted, and announce large costs for the purpose of reducing their individual cost. We present an approach that controls the detrimental effect that greedy users have upon the total cost.

3.6.1   Sensitivity to Approximate Costs

We begin by considering that the true user costs are $\hat{w}_j \in [1, \hat{B}]$, where $\hat{B}$ is a suitable upper bound placed on the exact costs. We suppose the costs that the users announce are derived by applying an operator T to $\hat{w}_j$, i.e. $w_j = T(\hat{w}_j)$. We define $\hat{W} = \sum_j \hat{w}_j$, and $\hat{p}_j = \hat{w}_j / \hat{W}$. If we build a code using p_j with lengths l_j, then the average length under $\hat{p}$ is $\sum_j \hat{p}_j l_j$. We show that if we design the code to minimize $\sum p_j l_j$, then we can design the operator T such that $\sum |\hat{p}_j l_j - p_j l_j|$ is small. Since $l_j \leq n$, we get $\sum_{j=1}^{n} |\hat{p}_j l_j - p_j l_j| \leq$

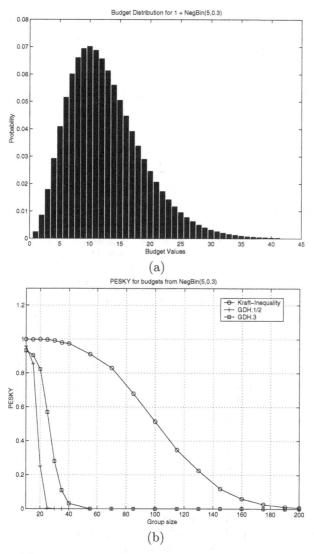

FIGURE 3.8. (a) Budget distribution, shifted version of a negative binomial distribution with parameters $s = 5$, and $p = 0.3$. (b) Corresponding PESKY

$n \left(\sum_{j=1}^{n} |\hat{p}_j - p_j| \right)$. We now derive a bound for $\sum_{j=1}^{n} |\hat{p}_j - p_j|$:

$$\sum_{j=1}^{n} |\hat{p}_j - p_j| = \sum_{j=1}^{n} \left| \frac{w_j}{W} - \frac{\hat{w}_j}{\hat{W}} \right| \tag{3.9}$$

$$\leq \frac{1}{W\hat{W}} \left[\sum_{j=1}^{n} |w_j(\hat{W} - W)| + \sum_{j=1}^{n} |W(w_j - \hat{w}_j)| \right] \tag{3.10}$$

$$\leq \frac{1}{W\hat{W}} \left[2W \sum_{j=1}^{n} |w_j - \hat{w}_j| \right] \tag{3.11}$$

$$= \frac{2}{\hat{W}} \left[\sum_{j=1}^{n} |w_j - \hat{w}_j| \right] . \tag{3.12}$$

We consider two cases for the operator T. The first case we consider is when T is a clipping operator, namely

$$T_{\hat{B}}(\hat{w}) = \begin{cases} \hat{w} & : \quad \hat{w} \leq \hat{B} \\ \hat{B} & : \quad \hat{w} > \hat{B} \end{cases}.$$

The clipping operator takes the true costs, $\hat{w}_j$, and leaves the value alone if it is in $[0, \hat{B}]$, otherwise it returns $\hat{B}$. Based on this, if we increase $\hat{B}$, then it is more likely that the true cost $\hat{w}_j$ will not be clipped. Consequently, it is clear that as $\hat{B} \to \infty$, we have more $w_j = T_{\hat{B}}(\hat{w}_j) = \hat{w}_j$, and thus the bound (3.12) tends to 0 as we increase $\hat{B}$. We shall examine the clipping operator later in this section. The second operation we consider is quantization. Here we consider the interval $[1, \hat{B}]$ divided into N equally sized quantization bins. The operator T then maps $\hat{w}$ to the nearest quantization value, and $|w_j - \hat{w}_j| \leq \hat{B}/(2N)$. In this case, we get

$$\sum_{j=1}^{n} |\hat{p}_j - p_j| \leq \frac{1}{\hat{W}} \left(\frac{\hat{B}n}{N} \right) \tag{3.13}$$

which tends to 0 as the number of quantization bins N increases. Therefore, in both the case of clipping and quantization, the parameters can be adjusted to bring the probability distribution p close to $\hat{p}$, and thus the designed average codelength $\sum p_j l_j$ close to the average codelength of using l_j under $\hat{p}$.

3.6.2 Sensitivity to Costs from Untrusty Users

We next consider the effect one user has upon the computational cost of the remaining users. In many scenarios, there may be a user that hurts the other users by either selfishly trying to make his cost small, or maliciously trying to make the total cost of the remaining users large. Recall that if the weights are ordered as $w_1 \geq w_2 \geq \cdots \geq w_n$ then the lengths of the Huffman code can be ordered as $l_1^* \leq l_2^* \leq \cdots \leq l_n^*$ [30]. Therefore, if a user would like to keep his cost as small as possible, he should announce as large of a weight as possible. Additionally, announcing a large weight causes the p_j of the other users to decrease, thereby increasing their codelengths (see [50] for the relationship between a symbol's codelength and its self-information). Thus, if a malicious user wishes to adversely affect the lengths of the other users, he should announce as large of a weight as possible.

We first derive the worst-case effect that one user can have upon the computational cost of the other group members when Huffman coding is used to construct the conference tree. We suppose that the malicious or selfish user is u_1, and that he publishes a large weight w_1. To determine how much extra cost does choosing a large w_1 impose upon the other $n-1$ users, we define $\breve{W} = \sum_{k=2}^{n} w_k$ and define the probability $q_j = w_j/\breve{W}$ for $j \in \{2, 3, \cdots, n\}$, and $q_1 = 0$. Then q_j represents the probabilities that would be used in constructing a conference tree if user u_1 were not participating. Let l_j^* denote the optimal codelengths constructed using p_j, and $\breve{l}_j^*$ be the optimal codelengths constructed using q_j. Since u_1 is not involved in the construction of $\breve{l}_j^*$, we have $\breve{l}_1^* = 0$.

We define the following quantities:

$$C^* = \sum_{j=1}^{n} w_j l_j^*, \quad \breve{C}^* = \sum_{j=2}^{n} w_j \breve{l}_j^*, \quad C_{ex}^* = \sum_{j=2}^{n} w_j l_j^*.$$

We are interested in comparing C_{ex}^*, which is the total cost of the remaining $n-1$ users given the probabilities p_j which incorporate u_1's cost, with $\breve{C}^*$, which is the total cost of the $n-1$ users $u_2, u_3, \cdots, u_n$ without considering u_1's announced cost.

Since $\breve{C}^*$ arises as the optimal code for the $n-1$ users with costs w_2, w_3, $\cdots$, w_n, we know $\breve{C}^*$ minimizes costs of the form $\sum_{j=2}^{n} w_j l_j$. In particular, C_{ex}^* must satisfy:

$$C_{ex}^* = \sum_{j=2}^{n} w_j l_j^* \geq \sum_{j=2}^{n} w_j \breve{l}_j^* = \breve{C}^*. \qquad (3.14)$$

We may derive an upper bound for C_{ex}^* by observing that the code given by $\breve{l}_j^*$ can be used to construct a code for p_j by taking $l_1 = 1$ and $l_j = \breve{l}_j^* + 1$. The optimal code for the weights $w_1, w_2, \cdots, w_n$ must be better than this code, and hence

$$C^* \leq w_1 + \sum_{j=2}^{n} w_j(\breve{l}_j^* + 1) = \breve{C}^* + W. \qquad (3.15)$$

Since $C_{ex}^* = C^* - w_1 l_1^*$, we have $C_{ex}^* \leq \breve{C}^* + W - w_1 l_1^* \leq \breve{C}^* + \breve{W}$. Gathering the results together, we get the overall bound $\breve{C}^* \leq C_{ex}^* \leq \breve{C}^* + \breve{W}$. The upper bound is achieved when $w_1 > \check{W}$, and hence, in the worst case, u_1 forces the other $n - 1$ users to spend an extra $\check{W}$ of resources.

Next, we consider the more general case where a fraction of the users are untrusty and announce large costs. Suppose that the true costs are $\hat{w}_j$, and that the announced costs are $\tilde{w}_j$. If the underlying statistics governing $\hat{w}_j$ are known, it is possible to determine which $\tilde{w}_j$ are outliers and remove those users from the group key formation procedure. However, in many cases, the value of the conference exists regardless of whether a few users were untrusty, and it is desirable to have those users in the conference. In this case, an approach must be used to reduce the detrimental effect of these bad users upon the cost of forming the entire group key.

We suppose that the CKA applies a clipping operator to the announced user costs $\tilde{w}_j$ to produce costs $w_j = T_B(\tilde{w}_j)$ that are used by the CKA in determining the conference tree. Ideally, we would like to build the conference tree using the exact costs $\hat{w}_j$, but these are not available. Instead, if the conference tree is built using w_j or $\tilde{w}_j$, the corresponding lengths l_j and $\tilde{l}_j$ are used with the exact costs $\hat{w}_j$, which can lead to an increase in the total cost.

To study the amount of additional cost incurred by using a code designed for w_j when the true costs are $\hat{w}_j$, we shall examine the average codelength. Hence we design codes for $p_j = w_j/W$ and $\tilde{p}_j = \tilde{w}_j/\tilde{W}$, where $\tilde{W} = \sum \tilde{w}_j$. We are interested in studying $\sum \hat{p}_j l_j$ and $\sum \hat{p}_j \tilde{l}_j$. The Kullback-Leibler divergence $D(\hat{p}\|p)$ describes the additional average codelength that different coding schemes incur when designed for the wrong distribution p when the correct distribution is $\hat{p}$ [30, 40, 50, 51]. Given a model distribution for the true user costs, the CKA can use $D(\hat{p}\|p)$ to determine the value of the clipping parameter B that minimizes the miscoding penalty.

We calculated the divergence for $n = 100$ users when the original costs w_j were drawn according to $10LN(0,1)+100$, where $LN(\mu,\sigma)$ is the lognormal distribution arising from a normal distribution with mean μ and variance σ. The lognormal distribution was chosen because it has a long tail. The probability that a user is untrusty was 0.05, and untrusty users were assumed to announce a cost $\tilde{w}_j = \hat{w}_j + Y$, where $Y = 1000$ and $w_j = T_B(\tilde{w}_j)$. The choice of $Y = 1000$ was arbitrary and chosen to represent a large bias that an untrusty user might place on his announced costs. An example divergence $D(\hat{p}\|p)$ for costs $\hat{w}_j$ drawn according to this distribution is presented in Figure 3.9. There is a minimum that appears at approximately $B^* = 150$. A system should be designed for the average case. In order to do this, the optimal clipping parameter should be averaged over many realizations of the costs. For costs drawn according to $10LN(0,1) + 100$, we averaged the optimal clipping value over 10000 realizations and found the mean optimal clipping value to be $\overline{B}^* = 150.44$ and the variance of the optimal clipping value as $\sigma_{B^*} = 25.60$.

The relative difference between the cost of using the Huffman-based conference tree using w_j and $\hat{w}_j$ are now compared. If $\hat{l}_j$ are the optimal code-

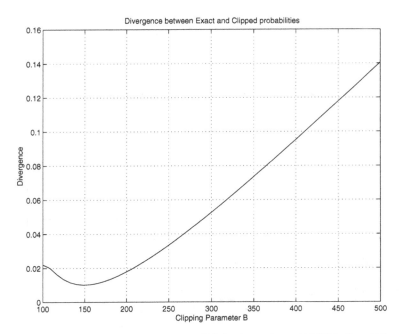

FIGURE 3.9. An example divergence $D(\hat{p}\|p)$ where $\hat{w}_j \sim 10LN(0,1) + 100$, and $w_j = T_B(\hat{w}_j)$.

lengths using $\hat{w}_j$, $\tilde{l}_j$ are the optimal codelengths constructed using $\tilde{w}_j$, and l_j are the optimal codelengths constructed using w_j, then we are interested in comparing

$$\rho = \frac{\sum_j \hat{w}_j l_j - \hat{w}_j \hat{l}_j}{\sum_j \hat{w}_j \hat{l}_j} \quad \text{and} \quad \tilde{\rho} = \frac{\sum_j \hat{w}_j l_j - \hat{w}_j \tilde{l}_j}{\sum_j \hat{w}_j \hat{l}_j}.$$

We calculated these values for the case when the exact costs were drawn according to $10LN(0,1) + 100$ with $Y = 1000$, while the probability of a user being untrusty was 0.05. The results were averaged over 100 realizations and are presented in Figure 3.10. The quantity ρ is presented for different clipping parameter values, and we observe that there is a range of minimal values from $B = 140$ to $B = 220$, which is roughly the region that the divergence curves predict. The clipped relative costs show a significant improvement over the unclipped relative costs. Without performing the clipping, the untrusty users force the entire group to spend an average of over 5% more than if the exact user costs were used. By performing the clipping operation, however, this detrimental effect can be significantly lessened to less than 0.5%.

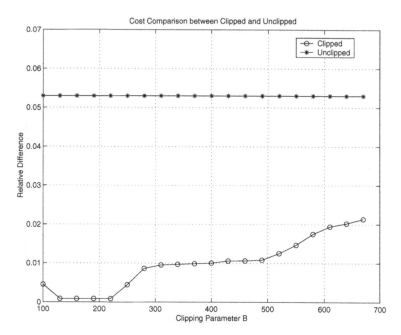

FIGURE 3.10. The relative costs ρ and $\tilde{\rho}$ are presented for when the exact user costs are drawn as $\hat{w}_j \sim 10LN(0,1) + 100$, and that there is an 0.05 likelihood that a user is untrusty, and $Y = 1000$.

3.7 Chapter Summary

In this chapter we have presented methods for establishing a conference key that are based upon the design of an underlying tree called the conference tree. In heterogeneous environments, where users have varying costs and budgets, the conference tree can be designed to address the user differences. We examined the design of the conference tree for three different cases. First, we studied the problem of minimizing the total cost of establishing the group key when the users had different costs. The problem of designing the conference tree was related to source coding, and techniques for designing source codes, such as Huffman coding, were employed to design the conference tree. The second case we investigated was when the users had the same cost, but different budget requirements. We observed that a necessary condition for a conference tree to exist for a given vector of budget requirements is that the budget vector satisfies the Kraft Inequality. We then presented a greedy algorithm that trimmed a feasible budget vector to achieve a length assignment that optimally reduces the total length of the conference tree. Finally, the third case we examined is when the users have both varying costs and budget requirements. We presented a computationally efficient near-optimal algorithm using a greedy incremental resource assignment strategy that achieves a total cost within 0.5% of the optimal solution for small group sizes.

We presented simulations comparing the total cost of the butterfly and Huffman-based schemes against the scheme of Ingemarsson et al., and the GDH family proposed by Steiner et al. Out of the class of non-centralized conference keying schemes, the Huffman scheme exhibited the least total cost. In situations where no single user has an extremely large budget, centralized conference keying schemes are unlikely to successfully establish a conference key. To investigate this phenomenon, we introduced the PESKY measure, which describes the probability that a conference keying scheme can establish a session key in the presence of budget constraints. We provided simulations where the user budgets were drawn according to different distributions, and in all cases the PESKY values for different group sizes were higher for our tree-based schemes than for either the GDH.1/2 or the GDH.3 schemes.

Next, we examined the effect that using false user costs would have on the total cost. It was shown that by increasing the quantization resolution, or by increasing the threshold level, that the difference between the total cost of using the exact and approximate costs for a given length assignment tends to 0. We then examined the effect a subset of users who falsely announce large costs has upon the total cost. In order to reduce the detrimental effect of designing a conference tree for falsely announced user costs, we proposed the use of a clipping operator to prevent untrusty users from being too greedy and minimize the divergence to determine the optimal threshold value. Simulations using the Huffman algorithm to construct the conference tree show that the optimal threshold values agree with those predicted by the divergence.

4
Optimizing Rekeying Costs in Group Key Agreement

The early design of contributory group key agreement schemes mostly focuses the efficiency of initial group key establishment, such as in [52–54]. These schemes, however, encounter high rekeying cost upon group membership changes. Later, Steiner et al. proposed a family of Group Diffie-Hellman (GDH) protocols by extending the two-party Diffie-Hellman (DH) protocols [55] to the group scenarios [56–58]. The GDH protocols achieve efficient key update upon user join, but still require high cost for member leave. Then, logical key tree structures are used to improve the scalability of contributory key agreements [59–61]. In tree-based contributory schemes, the group key can be updated by performing $\log n$ rounds of two-party DH upon any single user join or leave, where n is the group size.

What is the lowest possible cost of contributory group key agreement schemes? The theoretical analysis in [62] indicates that for any tree-based contributory group key management scheme, the lower bound of the worst case cost is $\Theta(\log n)$ rounds two-party DH for either user addition or deletion. That is, either the cost for adding a user or the cost for deleting a user is no less than $\Theta(\log n)$. In addition, it is obvious that at least one round of two-party DH needs to be performed for adding or deleting a user in any circumstance. Therefore, lowest possible cost for contributory key agreement is $\Theta(\log n)$ for user join and $O(1)$ for user leave; or $\Theta(\log n)$ for user leave and $O(1)$ for user join.

In this chapter, we describe two contributory schemes that employ novel tree structures and rekeying algorithms, with the aim to achieve the low bound of rekeying cost. Particular, the first method, referred to as JET [60, 63], uses a special join-tree/exit-tree topology and takes advantage of cost amortization. This method can significantly reduce the rekeying cost for user join. The second method is consist of a new key tree structure, called PFMH, and a key agreement protocol, called PACK [64, 65]. This method only needs $O(1)$ rounds of two-party DH upon any single user join event and $O(\log n)$ rounds of two-party DH upon any single user leave event, which achieves the lower bound described in the previous paragraph.

4.1 Join-Exit Tree for Reducing Latency in Key Agreement Protocols

The JET scheme described in [60] represents an important effort to reduce the cost in contributory key agreement. This scheme focuses on time efficiency, which is measured by the processing time in group key establishment and update. In order to participate in the group communications, a joining user has to wait until the group keys are updated. Since computing cryptographic primitives and exchanging rekeying messages are time-consuming, such waiting time is not negligible. Similarly, the amount of time needed to recompute a new group key reflects the latency in user revocation. Thus from a quality of service (QoS) perspective, the rekeying time cost is directly related to users' satisfaction and a system's performance.

The basic idea in JET is to employ a new key tree topology that has two small subtrees, the join and exit subtrees, located close to the root of the key tree. The sizes of join and exit trees should be at the log scale of the group size. With proper algorithms that handle the key update for join and leave events, an average asymptotic time cost for a join event is $O(\log(\log n))$, and also $O(\log(\log n))$ for a departure event when group dynamics are known *a priori*.

4.1.1 Time-efficiency Measurement

The time efficiency of DH-based contributory group key agreement is usually evaluated by the number of rounds needed to perform the protocol during a key update [52, 56, 66, 67]. However, in some schemes, the number of operations may be different in distinct rounds. For example, in GDH.2 [56], i modular exponentiations are performed in the i-th round. To address this problem, the notion of "simple round" was introduced in [68], where every party can send and receive at most one message in each round. In this section, we apply the notion of simple round in the tree-based contributory schemes. In each round, each user can perform at most one two-party DH operation. With the definition of simple round, the performance metrics for time efficiency are listed as below.

Average Join/Leave Time *User join time* is defined as the number of rounds to process key updates for a user join event. The average user join time, denoted by T_{join}, is defined as

$$T_{join} = \frac{R_{join}}{N_{join}}, \tag{4.1}$$

where R_{join} is the total number of DH rounds performed for N_{join} join events. Similarly, the *user leave time* is defined as the number of rounds to process key updates for a user leave event. The average user leave time, denoted by T_{leave}, is defined as

$$T_{leave} = \frac{R_{leave}}{N_{leave}}, \tag{4.2}$$

where R_{leave} is the total number of DH rounds performed for N_{leave} leave events. Let $N = N_{join} + N_{leave}$ and $R = R_{join} + R_{leave}$. The overall average processing time T is defined as

$$T = \frac{R}{N}, \tag{4.3}$$

where T can also be interpreted as a weighted average of T_{join} and T_{leave} as $T = \frac{N_{join}}{N}T_{join} + \frac{N_{leave}}{N}T_{leave}$.

4.1.2 Join-Exit Tree (JET) Topology

In this section, we describe the join-exit tree (JET) topology and and the associated key agreement algorithms.

As shown in Fig. 4.1(a), the join-exit tree consists of three parts: the *join tree*, the *exit tree*, and the *main tree*. It is a binary tree built upon the two-party DH protocol. In this section, the key tree that has the only main tree structure is referred to as the *simple key tree*.

The prior works have shown that, if a user joins the group at a location closer to the tree root, fewer number of keys need to be updated, thus the join time will be shorter. Similar reasoning applies to user departures. So the join tree and exit trees should be much smaller than the main tree. We define the *join tree capacity* and the *exit tree capacity*, denoted by C_J and C_E, as the maximum number of users that can be accommodated in the join and exit tree, respectively. The number of users in the join tree and the main tree are denoted by N_J and N_M, respectively.

In the JET scheme, a joining user will first be added to the join tree. Later on, when the join tree reaches its capacity, all users in the join tree will be relocated together into the main tree. In addition, when users' departure time is known, users who are most likely to leave in the near future will be moved in batch from the main tree to the exit tree. The design rationale of the join and exit trees resembles that of memory hierarchy in computer design [69]. Furthermore, the capacities of the join and exit trees can change

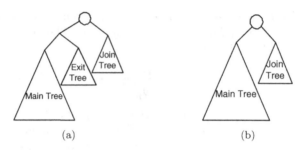

(a)   (b)

FIGURE 4.1. Topology for the proposed join-exit tree (a) join, exit, and main tree. (b) join and main tree.

over time, resulting in a dynamic key tree structure. For example, when there is no user in the exit tree, the key tree reduces to a main tree and join tree topology, as shown in Fig. 4.1(b).

4.1.3 The Join Tree Algorithm

The join tree algorithm consists of four parts: the join tree activation, the insertion strategy, the relocation strategy, and the join tree capacity update. When the group has only a few members, the join tree is not activated. As the group size increases and exceeds a threshold we activate the join tree and choose an initial join tree capacity. Such a threshold condition is referred to as the *activation condition* for the join tree. After the activation, any user joining the group is first inserted to a node in the join tree. The insertion node is chosen according to the *insertion strategy*. When the join tree is full, the members in the join tree are merged into the leaf nodes of the main tree. Such a process is called the *batch relocation*. Since the number of users in the main tree is changed after the batch relocation, the join tree capacity is updated according to a rule that relates the join tree capacity to the main tree user number. According to this rule, the *optimal join tree capacity* in the sense of time efficiency can be computed. We explain these four parts in details below.

User Insertion in the Join Tree

When the join tree is empty and a new user wants to join, the root of the current key tree is chosen as the insertion node. The insertion is done by treating the entire existing group as one logical user, and performing a two-party DH between this logical user and the new user. This process is illustrated in Fig. 4.2, where the new user M_5 becomes node 9, the root of the join tree. Member M_5 is paired up with the original root of the key tree (node 1) to perform a DH key exchange and the new group key is established as node 8. When the join tree is not empty, the insertion node is determined by Algorithm 6, where $usernumber(x)$ returns the number of users under a given node x in the key tree. After the insertion node is

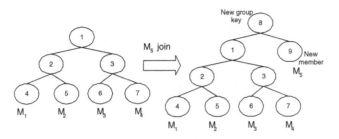

FIGURE 4.2. User join at the join tree root. Note that the new user M_5 becomes the root of the join tree.

found, the new member node performs a two-party DH key exchange with the insertion node. Then the keys on the path from the insertion node to the tree root are updated through a series of DH key exchange. Fig.4.3 illustrates the growth of the join tree from one user to eight users using the insertion strategy.

$x \leftarrow$ *join-tree-root* ;
while *usernumber*$(x) \neq 2^k$ *for some integer k* **do**
    $x \leftarrow rightchild(x)$;
end
insertion-node $\leftarrow x$

Algorithm 6: Finding the insertion node

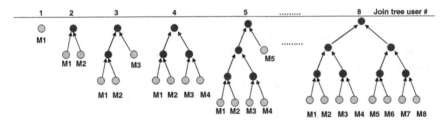

FIGURE 4.3. Sequential user join strategy (only the join tree is shown).

The Batch Relocation

There are two relocation methods that differ in whether the subgroup keys in the join tree are preserved. In the first method, all users in the join tree are viewed as a logical user during relocation, and this logical user is inserted into the shortest-depth leaf node of the main tree. Thus, the subgroup keys among the users in the join tree are preserved. This process is shown in Fig.4.4(a). Then all keys along the path from the insertion node to the

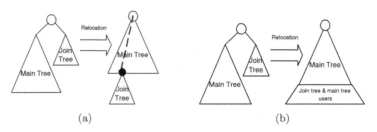

(a)   (b)

FIGURE 4.4. Relocation methods for the join tree (a) Method 1. (b) Method 2.

TABLE 4.1. Latency of Sequential User Join

k	1	2	3	4	5	6	7	8	9	10	...
$r(k)$	1	2	2	3	2	3	3	4	2	3	...

tree root are updated, which is indicated by the dash line in Fig. 4.4(a).
The reason to choose the shortest branch leaf node in the main tree as
the insertion node is to guarantee that the relocation time is at most the
log of the main tree size ($\lceil \log N_M \rceil$), because the shortest branch must be
smaller or equal to the average length of the branches, which is $\lceil \log N_M \rceil$.
The only exception comes when the main tree is a complete balanced tree,
the relocation time is $\log N_M + 1$, because one more level of the key tree
must be created to accommodate the new logical user.

In the second relocation method, we find the N_J shortest-depth leaf nodes
in the main tree as the insertion nodes for N_J join tree user. These insertion
nodes are found so that the unbalance-ness of the key tree can be alleviated
by the relocation process. Then we relocate the join tree users simultaneously to the insertion nodes. The keys on the branches from all original join
tree users to the tree root are updated in parallel and finally a new group
key is obtained. This process is illustrated in Fig.4.4(b). To analyze the
time complexity, we note that this relocation may fill up the empty nodes
at the shortest-depth leaf nodes of the main tree. The maximum depth of
any relocation path would not exceed $\lceil \log(N_M + N_J) \rceil$. Since the join tree
is much smaller than the main tree, the relocation time is upper bounded
by $\lceil \log N_M \rceil + 1$.

Although the two relocation methods have similar time complexity, the
first method will generally produce a skewed main tree. Since users may
leave from a branch longer than the average depth of the key tree, an
unbalanced key tree may cause the user departure time to be longer than
the case when a balanced key tree is used. The second relocation method
helps maintain the balance of the key tree, which reduces the expected cost
of leave events [66]. We shall choose the second relocation method in this
work because it takes into consideration both the join and leave time cost.

The Optimal Join Tree Capacity

Using the proposed insertion strategy, the user join latency for the k-th user in the join tree is measured as $r(k)$ rounds, which is listed in Table 4.1. We observe a special property of the sequence $r(k)$, namely,

$$r(2^p + q) = 1 + r(q), \quad p \ge 0, \ 0 < q \le 2^p, \tag{4.4}$$

where p is a non-negative integer, and q a positive integer. For the user join latency $r(k)$ in (4.4), the following inequality holds for any positive integer n, and equality is achieved when n is of power of 2:

$$\frac{1}{n} \sum_{k=1}^{n} r(k) \le \frac{1}{2} \log n + 1. \tag{4.5}$$

The proof is as follows.
Proof:
   We first use induction to show that when $n = 2^p$, $p = 0, 1, 2, ...$, the equality holds.
   When $n = 1$, Left_hand_side(LHS) = Right_hand_side(RHS) = 1.
   Next, we assume the equality holds for $n = 2^p$, namely,

$$\frac{1}{2^p} \sum_{k=1}^{2^p} r(k) = \frac{1}{2} \log 2^p + 1. \tag{4.6}$$

Consider the case of $n = 2^{p+1}$.

$$
\begin{aligned}
LHS &= \frac{1}{2^{p+1}} \sum_{k=1}^{2^{p+1}} r(k) \\
&= \frac{1}{2^{p+1}} \left(\sum_{k=1}^{2^p} r(k) + \sum_{k=1}^{2^p} (r(k) + 1) \right) \\
&= \frac{1}{2^{p+1}} \left(2 \cdot (\frac{1}{2} \log 2^p + 1) 2^p + 2^p \right) \\
&= \frac{1}{2} \log 2^{p+1} + 1 = RHS,
\end{aligned}
\tag{4.7}
$$

where (4.7) is obtained using the induction assumption (4.6).
   We now prove the inequality for any positive integer n. It is obvious to see that inequality is true for $n = 1, 2$. By induction, suppose that the inequality is true for all $1 \le n < 2^p + q$, and we consider $n = 2^p + q$, where $0 < q \le 2^p$.

$$LHS = \frac{1}{n} \sum_{k=1}^{n} r(k)$$

$$= \frac{1}{n}\left(\sum_{k=1}^{2^p} r(k) + \sum_{k=1}^{q}(r(k)+1)\right)$$

$$\leq \frac{1}{n}[(\frac{1}{2}\log 2^p + 1)2^p + q(\frac{1}{2}\log q + 1) + q] \tag{4.8}$$

$$= \frac{1}{2}\left\{\frac{1}{n}(2^p \log 2^p + q \log q + 2q)\right\} + 1, \tag{4.9}$$

where (4.8) is obtained using the induction assumption.

To prove that $(4.9) \leq \frac{1}{2}\log n + 1$ is equivalent to prove

$$\frac{2^p}{n}\log 2^p + \frac{q}{n}\log(4q) \leq \log n. \tag{4.10}$$

Applying the identity $\log k = \log e \cdot \ln k$ and $\ln k = \int_1^k \frac{1}{x}dx$, (4.10) can be written as an integration form

$$\log e\left\{\frac{2^p}{n}\int_1^{2^p}\frac{1}{x}dx + \frac{q}{n}\int_1^{4q}\frac{1}{x}dx\right\} \leq \log e \int_1^n \frac{1}{x}dx$$

$$\Leftrightarrow 2^p \int_{2^p}^n \frac{1}{x}dx + q\left[\int_1^n\frac{1}{x}dx - \int_1^{4q}\frac{1}{x}dx\right] \geq 0 \tag{4.11}$$

We denote $B = 2^p$ and fix p (hence B is fixed). Thus $n = B + q$. It is straightforward to see that (4.11) holds when $B + q \geq 4q$, or $1 \leq q \leq \frac{B}{3}$.

When $B/3 \leq q \leq B$, (4.11) is equivalent to

$$\frac{2^p}{n}\int_{2^p}^n \frac{1}{x}dx - \frac{q}{n}\int_n^{4q}\frac{1}{x}dx \geq 0. \tag{4.12}$$

Since q is the only variable in (4.12), let $f(q)$ be the LHS of (4.12), and consider $f(q)$ as a continuous function of q

$$f(q) = \frac{B}{B+q}\int_B^{B+q}\frac{1}{x}dx - \frac{q}{B+q}\int_{B+q}^{4q}\frac{1}{x}dx,$$

where $q \in [B/3, B]$. Taking the derivative of $f(q)$, we get

$$\frac{d}{dq}f(q) = -\frac{B}{(B+q)^2}\int_B^{4q}\frac{1}{x}dx < 0. \tag{4.13}$$

In this proof, we have showed that the equality holds when n is power of 2, i.e. $f(B) = 0$. We also showed that $f(q) > 0$ for $1 \leq q \leq \frac{B}{3}$. Since $f(B/3) > 0$, $f(B) = 0$, $f(q)$ is continuous on $[B/3, B]$ and $f'(q) < 0$, we must have $f(q) > 0$ on $[B/3, B]$. Thus (4.11) also holds for $B/3 \leq q \leq B$. This completes the proof.

End of Proof

Consider the average join time for x users joining the group starting form an empty join tree. These x users are inserted into the join tree one by one, then they are relocated together into the main tree. From previous analysis we can see that, when the main tree has N_M users, the average join tree relocation time is $\log N_M$, where we relax the integer value of the tree height to a continuous value to simplify analysis. Taking into account the relocation time, the average join time for these x users is

$$T_{join} = \frac{1}{x}(\sum_{k=1}^{x} r(k) + \log N_M). \tag{4.14}$$

Using (4.5), one can obtain

$$T_{join} \le \frac{1}{2}\log x + \frac{1}{x}\log N_M + 1. \tag{4.15}$$

Since it is not easy to minimize T_{join} directly, we try to minimize its upper bound over x. The optimal join tree capacity C_J that minimizes the upper bound is given by

$$
\begin{aligned}
C_J &= \arg\min_{x>0}\{\frac{1}{2}\log x + \frac{1}{x}\log N_M + 1\} \\
&= 2\ln N_M
\end{aligned}
\tag{4.16}
$$

The above analysis shows that, for a given number of main tree users N_M and the insertion rule specified by Algorithm 6, the optimal join tree capacity C_J is $2\ln N_M$. Since between two consecutive join tree relocations, the main tree size is fixed at N_M, the join tree capacity should also be fixed during this time at $C_J \approx 2\ln N_M$ and the average join time is upper bounded by

$$T_{join} \le \frac{1}{2}\log\log N_M + \frac{3}{2} + \frac{1}{2\ln 2} - \frac{1}{2}\log\log e. \tag{4.17}$$

This upper bound indicates that on average, a user needs to spend only $O(\log(\log n))$ rounds for a rekeying operation in user join, where n is the group size. It is noted that this asymptotic performance is not affected by the variation of the relocation time, because the relocation time of around $\log N_M$ rounds is averaged over $\log N_M$ join events, contributing approximately only one round to the average join cost. This validates the use of the approximate average relocation time $\log N_M$ in the above analysis.

For the joining users, since they can start to communicate once they are inserted into the join tree, their waiting time do not include the relocation time of $\log N_M$ rounds. The waiting time for the joining users is referred to as *user join latency*. One can see that the average user join latency, L_{join}, is also upper bounded as

$$L_{join} \le \frac{1}{2}\log(\log N_M) - \frac{1}{2}\log\log e + \frac{3}{2}.$$

The Join Tree Activation

To decide whether to activate the join tree, one should compare the average join time with and without employing the join tree. For a key tree structure with join tree, adding each user in the join tree incurs at most a time cost of $\log C_J$ rounds. Consider the average user join time for C_J users when the join tree changes from empty to full, followed by a batch relocation of $\log N_M$ rounds. The average join time for these C_J users satisfies

$$T_{join} \leq \log C_J + (\log N_M)/C_J. \tag{4.18}$$

If a simple key tree with only a main tree is used, the average join time would be at least $\log N_M$. Consequently, a reduction in time cost can be obtained by using the join tree when the following inequality holds,

$$\log C_J + (\log N_M)/C_J \leq \log N_M,$$

or equivalently,

$$\log N_M \geq \frac{C_J}{C_J - 1} \log C_J. \tag{4.19}$$

When the number of users in the group is large enough, a join tree should be activated to reduce the average join time. It can be proved that when $C_J = 2 \ln N_M$, the inequality (4.19) is satisfied for any $N_M > 8$. Thus a reasonable group size threshold is $TH_{join} = 8$. When the group size is smaller than or equal to 8, a simple key tree is used. Otherwise, the join tree is activated.

4.1.4   The Exit Tree Algorithm

In some group applications, users can estimate the duration of their staying time according to their own schedule. Such information can help reduce the time cost of rekeying operations in user departure. In the following analysis, we assume that we can obtain accurate information about users' duration of stay. In later sections, the cases of inaccurate or unavailable staying time will be discussed.

Similar to the join tree algorithm, the exit tree algorithm consists of four parts, namely, the activation condition, the batch movement, the user insertion in the exit tree, and the optimization of the exit tree capacity.

The Batch Movement

The *batch movement* refers to the operations to move the users that are likely to leave in the near future from the main tree to the exit tree. The group communications is not interrupted since the old group key can still be used before the batch movement is completed.

A batch movement takes place when there is a user leaving from the exit tree and a batch movement condition is satisfied. Denoting the number of users in the exit tree after the last batch movement as U_p, and the current number of users in the exit tree as U_c, we propose a batch movement condition as

$$U_c \leq \rho U_p, \tag{4.20}$$

where $\rho \in [0,1)$ is the *exit tree residual rate* (residual rate for short), a pre-determined parameter to control the timing of batch movement. In a batch movement, the first B users who are most likely to leave soon are moved to the exit tree, where B is referred to as the *batch movement size*. Starting from an empty exit tree ($U_p = 0$), the number of users in the exit tree after the k-th batch movement will be upper bounded by $\sum_{i=0}^{k-1} \rho^i B$. As k goes to infinity, the number of users in the exit tree converges to the upper bound $B/(1 - \rho)$. Therefore the exit tree capacity C_E is related to the batch movement size by

$$C_E = B/(1 - \rho). \tag{4.21}$$

A priority queue [70] can be used to keep the departure time of all the users in the main tree. This queue is referred to as the *leaving queue*. The users' departure time is obtained from their arrival time and their estimated staying time. The leaving queue will be update under two circumstances. First, after a batch relocation of the join tree, the departure information of the join tree users are added to the leaving queue. Second, after the batch movement of the exit tree, the departure information of the moved users are removed from the leaving queue.

User Insertion in the Exit Tree

The insertion locations for the users being moved into the exit tree are chosen to maintain the balance of the exit tree. For each user insertion, the leaf node with the minimum depth in the exit tree is chosen as the insertion node.

Optimal Exit Tree Capacity

Here we derive the optimal exit tree capacity that minimizes an upper bound of the average leaving time. Suppose that b users are moved together into the exit tree. A batch movement of these b users will incur a time cost of $(\log N_M + 2)$, where $\log N_M$ is the average height of the main tree, and the addition of 2 refers to the additional two levels above the main tree due to the use of the join tree and the exit tree (refer to Fig. 4.1(a)). If the exit tree capacity is x, each user leaving from the exit tree will incur at most a time cost of $(\log x + 2)$. Thus the average user leave time for these b users is bounded by

$$T_{leave} \leq \frac{1}{b}(\log N_M + 2) + (\log x + 2). \tag{4.22}$$

Using (4.21), $b = x(1-\rho)$, and minimizing the right hand side of (4.22) we obtain

$$
\begin{aligned}
C_E &= \arg\min_x \left\{ \frac{1}{(1-\rho)x}(\log N_M + 2) + (\log x + 2) \right\} \\
&= \frac{\ln N_M + 2\ln 2}{(1-\rho)}.
\end{aligned}
\tag{4.23}
$$

When the capacity of the exit tree is computed as in (4.23), the average leave time is bounded by

$$
T_{leave} \leq \log(\log N_M + 2) + \delta,
\tag{4.24}
$$

where $\delta = 2 - \log(1-\rho) + \log e - \log\log e$. Combining (4.23) and (4.21), we have

$$
B = \ln N_M + 2\ln 2.
\tag{4.25}
$$

A few comments should be made to provide more insights from the above analysis. First, the batch movement size B is only determined by the number of users in the main tree, and independent of the residue rate ρ. Second, there are actually only two parameters, B and ρ, in our system, since the exit tree capacity is a function of B and ρ as in (4.21). Third, with perfect departure information, the average leave time is bounded by $O(\log\log n)$, where n is the group size, and the residue rate ρ should be set to 0 to minimize the upper bound in (4.24). However, in practice, the choice of ρ is a tradeoff. When ρ is 0, a batch movement cannot be performed unless the exit tree is completely vacant. If some users inaccurately estimate their departure time and stay in the exit tree for a long period of time, no other users can utilize the exit tree during that period. When ρ is close to 1, batch movements are frequently performed, resulting in a large overhead. Based on experimental heuristics, we suggest setting ρ to around 0.5.

The Activation of Exit Tree

The average leave time using a simple key tree with N_M users is $\log N_M$. Comparing this result with the upper bound in (4.22), a reduction in the average leave time can be obtained if

$$
\frac{1}{(1-\rho)C_E}(\log N_M + 2) + (\log C_E + 2) \leq \log N_M.
\tag{4.26}
$$

Using (4.23), we simplify the above condition as

$$
\log N_M \geq \log C_E + \log e + 2.
\tag{4.27}
$$

Similar to the case of the join tree activation, it can be proved that when the exit tree capacity is chosen as in (4.23), the inequality (4.27) is satisfied for any $N_M > 256$. Thus we have found a threshold group size $TH_{leave} = 256$. When the group size is larger than this threshold, activating the exit tree can reduce the average leave time.

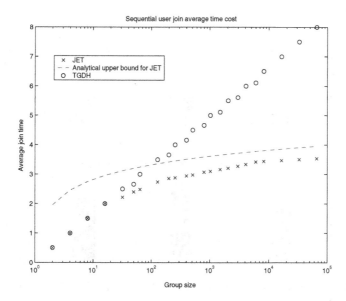

FIGURE 4.5. Average time cost for sequential user join.

4.1.5 Performance Analysis

In this section, three simulation experiments are described. The first simulation focuses on group key establishment, in which sequential user join is considered. The second and third simulation have both join and departure activities. In each simulation, the performance of JET is compared with that of TGDH scheme [66].

Key Establishment for Sequential User Join

For sequential user join, the JET protocol uses a simple key tree for small group size, and activates the join tree when the group size is larger than 8. The exit tree will not be activated. We compare the average join time for sequential user join using the proposed JET and TGDH [66] in Fig. 4.5. It can be seen that JET achieves the same performance as TGDH when the group size is small, and outperforms TGDH when the group size becomes large. Regarding the asymptotic performance, TGDH achieves an average time cost of $O(\log n)$, while the proposed JET scheme achieves $O(\log(\log n))$. The dashed line in Fig. 4.5 shows the theoretical upper bound for the average time cost from (4.17).

Experiment Using MBone User Activity Data

In this experiment, three user activity log files are chosen from three Multicast Backbone (MBone) multicast sessions [71] to describe user dynamics.

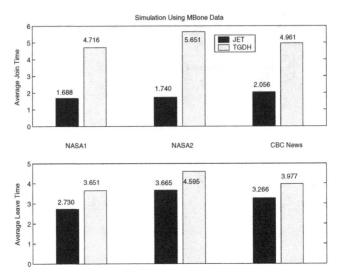

FIGURE 4.6. Average join and leave time for simulations using MBone data.

Two of these three sessions are NASA space shuttle coverage and the other one is CBC News World online test [1].

Fig. 4.6 shows the experimental results using JET and TGDH scheme. In the figure, JET has about 50% improvement over TGDH in user join, and about 20% improvement in user departure. It is worth noting that the improvement in user departure is not resulted from the use of the exit tree, since all the three sessions have maximum group size below 100 and the exit tree is not activated. From the study of the MBone multicast sessions, Ammeroth *et al.* observed that the MBone multicast group size is usually small (typically 100-200), and users either stay in the group for a short period of time or a very long time [72] [73]. Using the proposed JET scheme, the exit tree will not be activated for a small group size. However, when a user stays in the group for only a short period of time, it is highly likely that this user joins and leaves the group in the join tree without getting to the main tree. Thus the use of the join tree reduces both the user join time and the user leave time.

Experiments Using Simulated User Activity Data

In this experiment, user activities is generated according to the probabilistic model suggested in [72]. The duration of simulation is 5000 time units and is

---

[1]The sources of these MBone sessions are: (1) NASA-space shuttle STS-80 coverage, video, starting time 11/14/1996, 16:14:09; (2) NASA-space shuttle STS-80 coverage, audio, starting time 12/4/1996, 10:54:49; (3) CBC Newsworld on-line test, audio, starting time 10/29/1996, 12:35:15.

TABLE 4.2. Statistical Parameters for User Behavior

Duration	0-199	200-499	500-4499	4500-5000
λ_i	7	5	2	1
m_i	2500	500	500	500
Characteristic	long stay	short stay		

divided into four non-overlapping segments, T_1 to T_4. In each time segment T_i, users' arrival time is modelled as a Poisson process with mean arrival rate λ_i and users' staying time follows an exponential distribution with mean value m_i. The values of λ_i and m_i are listed in Table 4.2. The initial group size is 0. The simulated user activities consist of about 12000 join and 10900 leave events. The maximum group size is approximately 2800 and the group size at the end of simulation is about 1100.

In practice, users' accurate staying time will not always be available. More in-depth study has show that regardless of the accuracy in EST, the join tree scheme can improve the time efficiency for join events. The overall operation time is not very sensitive to the accuracy in EST. This is because inaccurate EST leads to user departures from the main tree. When users leave from the main tree, the JET scheme simultaneously relocates the users from the join tree to the main tree. As such, part of the join tree relocation cost can be amortized by the leave cost.

4.2 Optimizing Rekeying Cost

4.2.1 Performance Metric Review

The JET scheme in the previous section focuses on reducing time-cost associated with key agreement. In this section, the communication cost and computation cost are put into the consideration.

Next, we briefly review the performance measures and the implementation cost of the DH protocol between two groups.

Group key management schemes must be able to adjust group secrets subsequent to membership changes, including *single user addition, single user deletion, group merge,* and *group partition* [59]. Single user addition (deletion) means that one user joins (leaves) the group. Group merge (partition) involves multiple users who join (leave) the group simultaneously. The security requirements with dynamic membership include *group key secrecy, forward secrecy, backward secrecy,* and *key independence* [59]. Group key secrecy, which is the most basic property, requires that it should be computationally infeasible for a passive adversary to discover any group key. Forward secrecy requires that a passive adversary who knows a contiguous subset of old group keys cannot discover subsequent group keys, while backward secrecy requires that a passive adversary who knows a contiguous subset group keys cannot discover preceding group keys. Key independence, which is the strongest property, requires that a passive adversary who knows a proper subset of group keys cannot discover any other group key. According to [59], key independence can be achieved when both forward secrecy and backward secrecy are achieved.

The overhead of group key agreement involves *computation cost*, *communication cost* and *time cost*. Since most of the existing contributory key agreement schemes use two-party DH protocol [55] as a basic building module, the computation cost comes mainly from the cryptographic primitives that are needed to perform two-party DH, such as modular exponentiation, and the communication cost comes from sending and receiving rekeying messages. The time cost is used to describe the latency in group key establishing and updating. In contributory group key agreement, by exploiting possible parallelism when performing group key establishing and updating, the time cost can be significantly reduced.

Next we introduce the implementation of two-group DH (two-party DH among two groups), which is the basic building module for most tree-based contributory group key agreement schemes. Let A and B denote two subgroups, where the users in A share a common group key K_A, and the users in B share a common group key K_B. Let $f(K)$ (which we refer to as the blinded key of key K) denote the modular exponentiation operation, that is

$$f(K) = g^K \bmod p, \qquad (4.28)$$

where g is the exponential base and p is the modular base. The two-group DH can be implemented as follows. Each subgroup elects one member as its delegate, which will compute and send its blinded subgroup key to all members of the other subgroup. Suppose that member A_1 is the delegate elected by the subgroup A, and member B_1 is the delegate elected by the subgroup B. To perform two-group DH between these two subgroups, A_1 and B_1 need to exchange the following keying messages: A_1 sends the blinded key $f(K_A)$ to all members of subgroup B, and B_1 sends the blinded key $f(K_B)$ to all members of subgroup A. Now each member in A or B then calculates the new group key K_{AB} as follows:

$$K_{AB} = (f(K_B))^{K_A} \bmod p = (f(K_A))^{K_B} \bmod p. \qquad (4.29)$$

In this implementation, each member needs at least one modular exponentiation operation to calculate the new group key. If a delegate does not know its own subgroup's blinded key, one extra modular exponentiation operation is also needed to calculate the blinded key. For the communication cost, each delegate needs to send a keying message to all the members in the other subgroup. In this paper, we use $C_{cast}(n, \ell)$ to denote the communication cost needed to send a message with length ℓ to n nodes, and use C_{me} to denote the computation cost of a modular exponentiation operation. Thus, for each round of two-group DH with the size of subgroups being n_1 and n_2 and the keying message length being ℓ, the communication cost is $C_{cast}(n_1, \ell) + C_{cast}(n_2, \ell)$, and the computation cost is no more than $(n_1 + n_2 + 2)C_{me}$.

It is worth noting that sending a message to n nodes can be implemented in many ways. It can either be implemented through multicast communications, which we refer to as multicast-n, or be implemented through unicast, which we refer to as unicast-n. In general, the communication cost of an multicast-n operation is not the same as the communication cost of a unicast-n operation. The former usually incurs less communication cost than the latter. Further, the gap between the communication cost of an multicast-n operation and the communication cost of a unicast-n operation may vary according to the underlying network architectures. For example, in wireless networks the gap is usually very obvious due to the broadcast nature of wireless media, while in wired networks without link-level multicast support, the gap is usually not that obvious.

In this chapter, when analyzing the communication cost of sending a message to n nodes, both terms (multicast-n and unicast-n) will be used. Although the communication cost of multicast-n_1 and multicast-n_2 with $n_1 \neq n_2$ are usually different, to simplify the illustration, we will not distinguish them. Let $C_{multicast}(\ell)$ denote the communication cost of an multicast-n operation, and let $C_{unicast}(\ell)$ denote the communication cost of a unicast-1 operation, where ℓ is the length of the message to be sent. Further, when performing two-group DH between two subgroups, only messages exchanged are their blinded keys. Since in general all blinded keys have the same length, without loss of generality, the message length ℓ will not be explicitly stated. Besides exchanging blinded keys, a user may also need to send messages to all of the group members when it wants to join or leave a group. Additionally, $C_{broadcast}(\ell)$ is used to denote the communication cost incurred by broadcasting a message with length ℓ to all group members.

4.2.2 PFMH Key Tree Structure and Basic Procedures

In tree-based contributory group key agreement schemes, keys are organized in a logical tree structure, referred to as the *key tree*. In a key tree, the root node represents the group key, leaf nodes represent members' private keys, and each intermediate node corresponds to a subgroup key shared by all the members (leaf nodes) under this node. The key of each non-leaf node is generated by performing two-party DH between the two subgroups represented by its two children where each child represents the subgroup including all the members (leaf nodes) under this node [59]. Since two-group DH is used, the key tree is binary. For each node in the key tree, *key-path* denotes the path from this node to the root, and *co-path* denotes the sequence of siblings of each node on its key-path. Fig. 4.7 shows a simple key tree example with 6 members, where M_i denotes the ith group member and (l, v) denotes the vth node at level l of the tree. For example, for member

M_2, its key-path is the sequence of nodes $\{(3,1),(2,0),(1,0),(0,0)\}$, and its co-path is the sequence of nodes $\{(3,0),(2,1),(1,1)\}$.

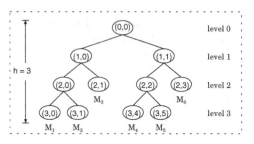

FIGURE 4.7. A simple key tree example

According to [59], in order to compute the group key, a node only needs to know its own key and all the blinded keys on its co-path. In other words, for a node being able to calculate the group key, it only needs to know its own keys and all the blinded keys on its co-path. For example, as shown in Fig. 4.7, M_2 only needs to know its own key and the blinded keys represented by the nodes $(3,0)$, $(2,1)$ and $(1,1)$ in order to calculate the group key.

A leaving user can leave from an arbitrary position in the key tree. In fact, for user leave, when group members have similar computation and communication capability, the best tree structure that reduces the worst-case rekeying overhead is a balanced key tree structure[2]. When using a balanced key tree structure, as in TGDH [59], the worst-case rekeying time cost for both user leave and user join is $O(\log n)$. In order to further reduce the rekeying time cost for user join, one way is to always insert the joining user at the root of the key tree, and consequently the rekeying time cost for single user join becomes $O(1)$. However, such scheme may result in an extremely unbalanced key tree structure and increase the rekeying cost for user leave to $O(n)$.

In order to achieve the lower bound for both user join and user leave simultaneously, a novel and efficient key tree structure is designed. This key tree structure is referred as the PFMH tree. PFMH tree is a combination of two special key tree structures: *partially-full* (PF) key tree and *maximum height* (MH) key tree.

In this chapter, the size of a key tree means the total number of leaf nodes in the tree, the function log() and $\log_2()$ will be used exchangeably, and when we say a "full (key) tree", we always mean a fully balanced binary (key) tree with size 2^k, where k is a non-negative integer.

Theorem 2 (PF key tree). *Let T be a binary key tree of size n, and let $n' = 2^{\lfloor \log n \rfloor}$. T is a PF key tree if and only if it satisfies one of the following*

---

[2]The case that users have varying computation and communication capabilities is not considered in this chapter.

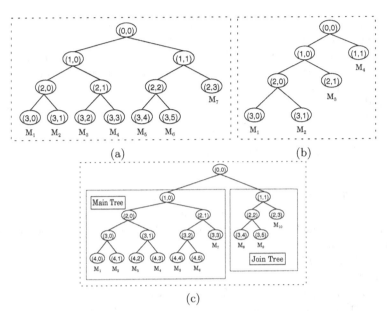

FIGURE 4.8. Some examples of PF/MH/PFMH key trees (a) PF key tree. (b) MH key tree. (c) PFMH key tree.

properties: 1) T is a full key tree; 2) the left subtree of T is a full key tree with size n', and the right subtree of T is a PF key tree with size $(n - n')$.

Theorem 3 (MH key tree). *A key tree T of size n is a MH key tree if and only if it satisfies one of the following properties: 1) $n = 1$, and T is a tree with only one leaf node; 2) the right subtree of T is a leaf node, and the left subtree of T is a MH key tree with size $n - 1$.*

Theorem 4 (PFMH key tree). *A key tree T of size n is a PFMH key tree if and only if it satisfies one of the following properties: 1) T is a PF key tree; 2) the left subtree of T is a PF tree, and the right subtree of T is a MH tree.*

According to the above definitions, one can see that the height of a PF key tree with size n is $\lceil \log n \rceil$, the height of a MH tree with size n is $n - 1$. Without introducing ambiguity, we will use $\lceil \log n \rceil$ and $\log n$ exchangeably in this chapter. Also, given a PFMH key tree T, let *main tree* refer to the PF subtree of T, denoted by T_{main}, and let *join tree* refer to the MH subtree of T, denoted by T_{join}. It is easy to see that the height of T_{main} is always bounded by $\log n$. Fig. 4.8 illustrates these special key tree structures. Next two basic procedures to manage and update PFMH key trees: *unite* and *split*, are described.

Let $\mathcal{T} = \{T_1, \ldots, T_L\}$ be a set of full key trees. Each key tree $T_i \in \mathcal{T}$ represents a subgroup, and each leaf node of T_i is a member of this subgroup. If a group member belongs to T_i and $T_i \in \mathcal{T}$, then this group member belongs to $\mathcal{T}$. The procedure *unite($\mathcal{T}$)* is to combine those key trees in $\mathcal{T}$ into a single PF key tree through performing a series of two-group DH among these subgroups as well as the subgroups generated during this procedure. In general, given a set of full key trees $\mathcal{T}$, the result of *unite($\mathcal{T}$)* may not be unique, but all of the obtained PF key trees have similar structure. In this paper we consider a special case where the full key trees in $\mathcal{T}$ are ordered and indexed according to their sizes. And, any group member in $\mathcal{T}$ knows the indices and sizes of any trees in $\mathcal{T}$ as well as these structure of these trees. The structure of a tree refers to the list of group members belonging to this tree and their exact positions in this tree. Then a group member can decide with whom it should perform two-group DH and in what order.

▷ $\mathcal{T} = \{T_1, \ldots, T_L\}$; $|T_i| \geq |T_j|$ for any $1 \leq i < j \leq L$; each member in $T_i \in \mathcal{T}$ knows the index and size of any tree $T_j \in \mathcal{T}$ as well as the structure T_j, including the list of group members in T_j and their exact positions in T_j.

$\mathcal{T}' = \mathcal{T}$; $L' = L$;

while *($|\mathcal{T}'| > 1$)* **do**

 for *(each pair of trees $T_i, T_{i+1} \in \mathcal{T}'$)* **do**

  if *((the total number of trees in $\mathcal{T}'$ with size equal to $|T_i|$ and with index before T_i is even) AND ($|T_i| = |T_{i+1}|$ or T_{i+1} is the tree in $\mathcal{T}'$ whose index is the largest))* **then**

   Two delegates will be elected by subgroups T_i and T_{i+1} to perform two-group DH between them, and a new group key K will be generated. A new key tree will be generated with its root node representing K, with the left child of the root node being T_i and with the right child of the root node being T_{i+1}. Remove T_i and T_{i+1} from $\mathcal{T}'$.

  end

 end

 Put all newly generated key trees in this round into $\mathcal{T}'$ and let L' be the total number of key trees now in $\mathcal{T}'$. Re-index all the key trees in $\mathcal{T}'$ with integers ranging from 1 to L in such a way that a tree is assigned index i (that is, this tree's name becomes T_i) if and only if condition 1 and condition 2 are satisfied. 1) for any tree $T_j \in \mathcal{T}'$ with index $j < i$, all subtrees of T_j that directly come from $\mathcal{T}$ have lower indices than all subtrees of T_i that directly come from $\mathcal{T}$ 2) for any tree $T_j \in \mathcal{T}'$ with $j > i$, all subtrees of T_j that directly come from $\mathcal{T}$ have higher indices than all subtree of T_i that directly come from $\mathcal{T}$.

end

Return the remaining tree T_1 in $\mathcal{T}'$, which is the final PF key tree. Meanwhile, each member in the final PF key tree construct the final key tree structure locally by following the above key tree generation procedure.

Algorithm 7: *unite($\{T_1, \ldots, T_L\}$)*

Procedure 7 presents one specific implementation of *unite(T)* for this special case. According to Procedure 7, the whole procedure is partitioned into many rounds. At the beginning of each round, there remains a set of full trees (subgroups) indexed according to their sizes and their subtrees' indices in previous rounds. The larger the size of a subgroup, the lower its index. In each round, a remaining subgroup may either keep alone or be paired with another remaining subgroup according to the following rule: two subgroups T_i and T_j $(i < j)$ will be paired together if and only if all of the three conditions can be satisfied:

- There is no other remaining subgroup in the round with index lying between i and j;

- The total number of subgroups with size equal to $|T_i|$ and with index lying before T_i is even;

- $|T_i| = |T_j|$ or T_j is the subgroup with largest index.

It is easy to see that in each round, a subgroup will either keep alone or be paired with one and only one other subgroup to build a larger subgroup. Further, in each round all pairs of subgroups can perform two-group DH between them in parallel, which can significantly reduce the time cost. If Procedure 7 is followed by all group members, the obtained PF key tree is unique and each member can know its location in the final PF before starting the procedure, and each member can locally construct the final PF tree without explicitly exchanging key tree updating information.

Given a key tree T, the procedure *split(T)* is to partition T into a set of full key trees with minimum set size. Specifically, after applying the procedure *split(T)*, any obtained key tree is a full key tree, and no any two or more obtained key trees comes from any full subtree of T. Procedure 8 presents a way to locally and virtually *split* a key tree, where "locally" means that no inter-communication is needed among group members and each member only needs to update the key tree structure maintained by itself locally, while "virtually" means that no any two-group DH is needed to perform "split". Meanwhile, the set of obtained full key trees are also indexed according their size and their positions in the original key tree.

Fig. 4.9 shows two examples of key tree update when applying *unite* and *split* procedures. The left figure demonstrates how the key tree is updated when 5 full key trees are *united* into a PF key tree. The right figure demonstrates how the key trees are updated when a PFMH key tree is *split* into a set of full key trees.

In the *split* procedure, each group member (leaf node) only needs to truncate the current key tree maintained by itself, so no communication cost and negligible computation and time cost are needed. In the *unite* procedure, extra cost will be incurred when performing a sequence of two-group DH to generate the new key tree.

---

if *(T is a full tree)* **then**
  Return $\{T\}$;
else if *(T is empty)* **then**
  Return $\emptyset$.
else
  Let T_{left} and T_{right} be the left and right subtrees of T; Return
  $split(T_{left}) \bigcup split(T_{right})$.
end
Let L be the number of obtained full key trees. Index these key trees
with the integers ranging from 1 to L in such a way that a tree is
indexed as T_i if and only if: 1) for any tree $T_j \in T'$ with $j < i$,
$|T_j| > |T_i|$ or T_j lies in the left side of T_i in T; and 2) for any tree
$T_j \in T'$ with $j > i$, $|T_j| < |T_i|$ or T_j lies in the right side of T_i in T.

Algorithm 8: $split(T)$

Next we analyze the cost associated with the *unite* procedure described
in Procedure 7. The results will be used later to analyze the cost of those
proposed key agreement protocols.

Theorem 1: Let $T = \{T_1, \ldots, T_L\}$ be a set of full key trees with
$\sum_{i=1}^{L} |T_i| = n$ and $|T_1| \geq |T_2| \geq \ldots \geq |T_L|$, and with the subscript ℓ being
the index of the full tree T_ℓ. Assume that any group member in any full tree
$T_i \in T$ knows the index and size of any tree $T_j \in T$ as well as the structure
of T_j. Then the costs associated with the $unite(T)$ using Procedure 7 can
be bounded as follows:

1) The time cost, which is the number of parallel rounds that needs to
   executed, is upper-bounded by $\log n$ in all situations.

2) The total communication cost is upper-bounded by $2(L-1)C_{multicast}$
   in all situations provided that the exchange of keying materials between two subgroups during performing two-group DH is implemented
   using multicast.

3) Consider the special situation that $|T_i| = 1$ for all $1 \leq i \leq L$, the total
   computation cost is upper-bounded by $n(\log n + 2)C_{me}$, and the total
   communication cost is upper-bounded by $(n \log n)C_{unicast}$ provided
   that the exchange of keying materials between two subgroups during
   performing two-group DH is implemented using unicast.

4) Consider the special situation that $|T_i| \neq |T_j|$ for any $1 \leq i \neq j \leq L$,
   the total computation cost is upper-bounded by $2(n + \log n)C_{me}$, and
   the total communication cost is upper-bounded by $2nC_{unicast}$ provided that the exchange of keying materials between two subgroups
   during performing two-group DH is implemented using unicast.

5) Consider the special situation that $|T_1| \geq n/2$ and for any tree $T_i \in T$
   there exists no more than one other tree in T with the same size as T_i,

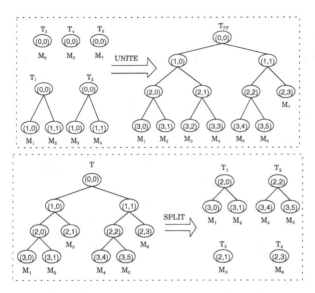

FIGURE 4.9. Examples of key tree update after applying *unite* and *split* procedures

the total computation cost is upper-bounded by $(2.5n + 2L)C_{me}$, and the total communication cost is upper-bounded by $2.5nC_{unicast}$ provided that the exchange of keying materials between two subgroups during performing two-group DH is implemented using unicast.

6) Consider the special situation that $|T_1| < n/2$ and for each tree $T_i \in \mathcal{T}$ there exists no more than one other tree in $\mathcal{T}$ with the same size as T_i, the total computation cost is upper-bounded by $(3n+2L)C_{me}$, and the total communication cost is upper-bounded by $3nC_{unicast}$ provided that the exchange of keying materials between two subgroups during performing two-group DH is implemented using unicast.

Proof of Theorem 1

1) Consider the worst-case scenario: $L = n$, that is, $|T_l| = 1$ for all l. Then Procedure 7 works as follows: in the first round, the set of group members are partitioned into $\lceil n/2 \rceil$ subgroups, with each subgroup consisting of 1 or 2 members. For any subgroup of size 2, two-group DH is performed between the two members in this subgroup to generate a new key tree of size 2. In the ith round, the set of existing key trees are partitioned into $\lceil n/2^i \rceil$ subgroups, with each group consisting of 1 or 2 existing key trees. If there is a subgroup consisting of only one existing key tree, then this key tree must have the minimum size (largest index) among all the existing trees. For any subgroup with two existing key trees, two-group DH is performed between these two key trees to generate a new key tree with its right child be the key tree which has smaller size (larger index). Repeat this procedure until only one tree is left, which is the final PF key tree. Since there are

only n group members, at most $\log n$ rounds are needed, so the time cost is upper-bounded by $\log n$. For other scenarios where there exists $|T_i| \neq 1$, the time cost is always no more than $\log n$, since in these cases T_i can be viewed as the result of merging all the leaf nodes in T_i without introducing any time cost.

2) Since we need and only need to perform $L - 1$ times of two-group DH protocols to unite L full key trees into one PF tree, and since each two-group DH protocol needs 2 multicast in communication cost provided that the exchange of keying material between two subgroups during performing two-group DH is implemented using multicast, the total communication cost is always upper-bounded by $2(L - 1)C_{multicast}$.

3) According to Procedure 7 we know that at most $\log n$ rounds of two-group DH need to be performed in this situation. At the first round each member calculates its blinded key and a new subgroup key. At i^{th} round $(i > 1)$, at most $\lceil n/2^{i-1} \rceil$ users (which are selected as delegates) need to calculate blinded keys and at most n users need to calculate their subgroup keys. Following this analysis one can see that the total computation cost is upper-bounded by $n(\log n + 2)C_{me}$. Further, if the exchange of keying material between two subgroups during performing two-group DH is implemented using unicast, it is easy to see that a blinded key needs to send to a certain member if and only if this member needs to calculate a key for a newly generated subgroup that it belongs to, which is equivalent to say that the total communication cost is upper-bounded by $(n \log n)C_{unicast}$ in this situation.

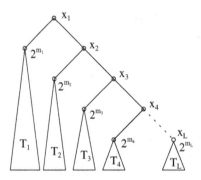

FIGURE 4.10. Obtained PF key tree after applying *unite* procedure

4) In this special situation, according to the definition of PF tree, it is easy to check that the key tree illustrated in Fig. 4.10 is the obtained PF key tree after applying Procedure 7. Assume that the size of each full subtree T_i is 2^{m_i}, and let x_i denote both the PF subtree and its size. According to Procedure 7, it is seen that the PF subtrees $x_L, \ldots, x_1$ are generated sequentially with x_L first (directly from T) and x_1 last. Also, when x_i is generated, at most $x_i + 2$ modular exponentiation operations are needed,

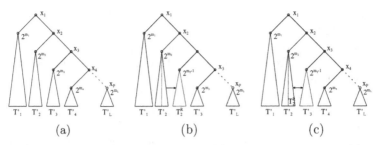

FIGURE 4.11. Analysis of computation cost. (a) Obtained PF key tree; (b) Create a virtual subtree for T_2^R; (c)Exchange T_2^R with T_3'.

so the total computation cost is upper-bounded by

$$C_{me} \sum_{i=1}^{L}(x_i + 2) = C_{me}\left(2\log n + \sum_{i=1}^{L} x_i\right). \tag{4.30}$$

Since

$$x_i = 2^{m_i} + x_{i+1}, \tag{4.31}$$

$$x_i \geq 2x_{i+1}, \tag{4.32}$$

$$\sum_{i=2}^{L} x_i < x_1 = n, \tag{4.33}$$

the above bound is further upper-bounded by $2(n + \log n)C_{me}$. Meanwhile, it is easy to check that the total communication cost is upper-bounded by $2nC_{unicast}$ in this situation provided that the exchange of keying material between two subgroups during performing two-group DH is implemented using unicast.

5) In this special situation, let the key tree illustrated in Fig. 4.11 (a) be the PF tree obtained after applying the *unite* procedure. Now consider the computation cost incurred by the full subtree T_i':

- Case 1: If T_i' comes directly from the original set T, the cost introduced by T_i' has been totally included in (4.30).

- Case 2: If T_i' is the merging result of two full trees directly from the original set T and each with size 2^{m_i-1}, compared with the first case, extra computation cost is needed to first merge the two trees into a single full tree. Since the total number of leaf nodes in T_i' is 2^{m_i}, and each leaf node needs 1 modular exponentiation to calculate the new subgroup key associated to T_i', the extra computation cost introduced by T_i' is $2^{m_i}C_{me} + 2$.

- Case 3: If T_i' is the merging result of more than two full key trees of the original set T, since we have assumed that for each size the number of trees with this size in T is no more than 2, then at least one child of T_i' with size 2^{m_i-1} comes directly from T. Let T_i^L and T_i^R

be the left and right child of T'_I, and assume that T^L_i comes directly from T. In this case, either there exists no key tree with size 2^{m_i-1} in the right side of T'_i, or if there exists $|T'_{i+1}| = 2^{m_i-1}$, then T'_{i+1} must come directly from T in order not to violate the assumption that no more than two key trees in T have the same size, and T'_{i+1} will not introduce extra cost except those included in (4.30). If there exists no subtree with size 2^{m_i-1} in the right side of T'_i, we add a virtual subtree T^R_i to the generated PF tree as in Fig. 4.11 (b) and move all the cost introduced by merging smaller full trees into this subtree. If $|T'_{i+1}| = 2^{m_i-1}$, we can simply exchange the subtree T'_{i+1} with the right subtree T^R_i of T'_i that is not directly from the original set T, as in Fig. 4.11 (c). Now the total cost is kept to be the same but the extra cost introduced by T'_i is the same as in case 2.

Following the above analysis and the condition that $|T_1| \geq n/2$ (that is, T_1 comes directly from T), the total extra computation cost that are not included in (4.30) is upper-bounded by $\sum_{i=1}^{m_2} 2^i C_{me}$. Now the total computation cost is upper-bounded by

$$2LC_{me} + C_{me}\left(\sum_{i=1}^{m_2} 2^i + \sum_{i=1}^{L} x_i\right) \qquad (4.34)$$

By applying (4.31), (4.32), (4.33) and $|T_1| = 2^{m_1} \geq n$, we have

$$\sum_{i=1}^{m_2} 2^i + \sum_{i=1}^{L} x_i \leq 2^{m_2+1} + 2x_1 \leq x_1/2 + 2x_1 = 2.5x_1 = 2.5n \qquad (4.35)$$

That is, the total computation cost is upper-bounded by $(2.5n + 2L)C_{me}$. Meanwhile, we can conclude that in this situation when the exchange of keying materials is implemented using unicast, the total communication cost is upper-bounded by $2.5nC_{unicast}$.

6) For the special situation that $|T_1| < n/2$ and for each tree $T_i \in T$ there exists no more than 1 other tree in T with the same size as T_i, by following the same analysis as in (5), we can show that the total computation cost is upper-bounded by

$$2LC_{me} + C_{me}\left(\sum_{i=1}^{m_1} 2^i + \sum_{i=1}^{L} x_i\right), \qquad (4.36)$$

where the only change from (4.34) to (4.36) is that m_2 is changed to m_1 due to the reason that T'_1 does not come directly from T.

By applying (4.31) one can derive that

$$\sum_{i=1}^{m_1} 2^i + \sum_{i=1}^{L} x_i = x_1 + (x_2 + 2^{m_1}) + \sum_{i=1}^{m_2} 2^i + \sum_{i=3}^{L} x_i \leq 2x_1 + 2^{m_2+1} + x_2 \leq 3x_1 = 3n$$

$$(4.37)$$

That is, the total computation cost is upper-bounded by $(3n + 2L)C_{me}$. Meanwhile, we can conclude that in this situation when the exchange of keying materials is implemented using unicast, the total communication cost is upper-bounded by $3nC_{unicast}$.
End of proof.

4.2.3   PACK: an PFMH tree-based contributory group key agreement

In this section we describe the proposed PFMH tree-based contributory group key agreement protocol suite, referred to as PACK. As a contributory scheme, in PACK each group member equally contributes its share to the group key and this share is never revealed to the others. To satisfy the security requirements, PACK includes a set of rekeying protocols to update the group key upon group membership change events. Compared with the existing tree-based contributory group key agreement schemes, PACK can achieve minimum rekeying time cost upon membership change events in the sense that for any single user join event, the rekeying time cost is of order $O(1)$, and for any single user leave event, the rekeying time cost is of order $O(\log n)$. Meanwhile, the rekeying computation and communication cost can also be significantly reduced compared with the existing tree-based contributory group key agreement schemes. This is achieved through adopting the proposed PFMH tree as the underlying key tree structure and introducing *phantom* nodes in the key tree to handle member leave.

In PACK, each member will maintain and update the global key tree locally. Each group member knows all the subgroup keys on its key-path, and knows the ID and the exact location of any other current group member in the key tree. As to be shown next, upon group membership change event, a group members only needs to update the global key tree maintained by itself, which can greatly reduce the communication overhead. In PACK, when a new user joins the group, it will always be attached to the root of the join tree to achieve $O(1)$ rekeying cost in terms of computation per user, time and communication. When a user leaves the current group, according to the leaving member's location in the key tree as well as whether this member has phantom location in the key tree, different procedures will be applied, and the basic idea is to update the group key in $O(\log n)$ rounds and simultaneously reduce the communication and computation cost.

Single User Join Protocol

When a prospective user M wants to join the group $\mathcal{G}$, it initiates the *single user join* protocol by broadcasting a request message that contains its member ID, a join request, its own blinded key, some necessary authentication information and its signature for this request message. After receiving this user join request message, the current group members will check whether

M has the privilege to join the group based on certain group access control policies. If M has the authorization to join, the key tree will be updated by incorporating M's share, and a new group key will be generated in order to incorporate a secret share from M and to guarantee group keys' backward secrecy. Procedure 9 describes the single user join protocol in PACK.

---

triangleright T is the PFMH key tree of group $\mathcal{G}$, T_{main} is the main tree of T, T_{join} is the join tree of T.

if *(T_{join} is empty)* **then**
> A delegate will be elected by group $\mathcal{G}$ to perform two-group DH with M, and a new group key K will be generated. A leaf node will be created to represent M and a new root node will be created to represent K with its right child being the node representing M and its left child being T_{main}. The node representing M becomes the join tree of the updated key tree.

else
> **Round 1**: A delegate will be elected by group T_{join} to perform two-group DH with M, and a new subgroup key K_{join} will be generated. A leaf node will be created to represent M and a new intermediate node will be created for K_{join} with its right child being M and its left child being the old T_{join} ;
> **Round 2**: Two delegates will be elected separately by T_{main} and the new join tree to perform two-group DH between them, and a new group key K will be generated. A new root node will be created to represent K with its right child being T_{join} and its left child being T_{main}.

end
Each current member updates the key tree maintained by itself locally according to the above key tree update procedure, and a delegate will send an updated copy of the key tree to the new joining member M.

Algorithm 9: $join(\mathcal{G}, M)$

In PACK, the rekeying upon single user join needs to perform at most 2 rounds of two-group DH. If the join tree is not empty, a new join tree is generated by performing two-group DH between the new member and the old join tree, with the left subtree being the old join tree and the right subtree being the node representing the new member. If the join tree is empty, the node representing the new member becomes the join tree. The group key is generated by performing two-group DH between the new join tree and the main tree. Since all the current members know the group key tree structure and knows the location that the new member should be put in, they can update the key tree themselves.

Fig. 4.12 shows two examples of key tree update upon single user join events. In the first example, the join tree is empty, and the main tree consists

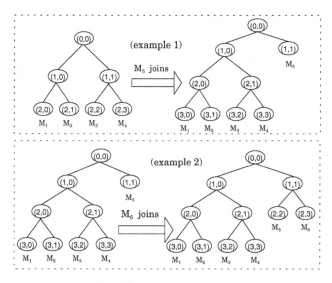

FIGURE 4.12. Examples of key tree update upon single user join event

of 4 members. After the new member M_5 joins the group, a new node is created to act as the new root, and the node $(1,1)$ becomes the new join tree which represents M_5. In the second example, when M_6 joins the group, at the first round, two-group DH is first performed between M_5 and M_6 to generate a new join tree, at the second round, two-group DH is performed between the new join tree and the main tree to generate a new group key.

TABLE 4.3. Rekeying cost upon single user join event

	communication cost in term of multicast	communication cost in term of unicast
case 1	$2C_{multicast}$	$nC_{unicast}$
case 2	$4C_{multicast}$	$(n + |T_{join}| + 1)C_{unicast}$
	time cost	computation cost
case 1	1	$(n + 2)C_{me}$
case 2	2	$(n + |T_{join}| + 3)C_{me}$

Table 4.3 lists the rekeying cost upon single user join event in PACK where n denotes the total number of leaf nodes in the new group and $|T_{join}|$ is the old join tree size. Case 1 considers the situation that the join tree is empty, and the protocol only needs to perform one round of two-group DH. Case 2 considers the situation that the join tree is not empty, and the protocol needs to perform two rounds of two-group DH. For case 2, the term $|T_{join}| + 2$ in the computation cost comes from performing two-group DH between the new member and the old join tree. Since in general $|T_{join}| \ll n$, this term usually can be ignored.

It is worth pointing out that when calculating the time complexity, we have not considered the extra time needed for the join user to tell the group that it wants to join. However, this does not affect the results because in the time complexity analysis, the "round" is used as unit. In other words, It is not strictly require that the two messages exchanges are synchronized. Instead, how this can be implemented is really depend on the specific implementation of the two-ground DH.

Single User Leave Protocol

When a current group member M wants to leave the group, it broadcasts a leave request message to initiate the single user leave protocol, which contains its ID, a leave request and a signature for this message. Once M leaves the group, the group key will be updated to remove M's share, and all the keys on M's key-path will be updated to maintain group keys' forward secrecy. In PACK, to reduce the rekeying cost upon single user leave event, the concept of *phantom* node is introduced. This concept allows an existing member to simultaneously occupy more than one leaf node in the key tree. In particular, when member M leaves the group, another group member M' will move to the position occupied by M in the key tree, generate a new secret key, and all the keys on M's key-path will be recursively updated. It is worth noting that here "moving" only means that each member adjusts the location of M' and M in the key tree. After moving M' to M's position, the node that M' previously occupied will not be deleted immediately. As a result, now M' occupies two leaf nodes in the key tree. We refer to the node associated to M''s previous position as the *phantom* node, which is known by all group members. In order to maintain group keys' forward secrecy, a phantom node should be deleted no later than the associated group member leaving the group. Procedure 10 describes the single user leave protocol in PACK.

SCENARIO I: This scenario considers the case that the leaving member M is in the join tree, and the size of the join tree is no larger than $\log n$. In this case, since the depth of the join tree is no more than $\log n$, one can simply remove M's share from the group key by removing M from the key tree, changing one current member's secret share (which member's share should be changed is described in Protocol 10), and recursively updating all the keys on M's key-path. Meanwhile, all members update the key tree maintained by themselves.

Let h be M's depth in T. Since at most $h - 1$ rounds of two-group DH protocols need to be performed recursively, the time cost is upper-bounded by $h - 1$. Except the last round which involves all the existing members, in $i^{th}(1 \le i < h - 1)$ round at most $|T_{join}| - h + i + 1$ members are involved. Then the total computation cost is upper-bounded by $(n+h-1+ \sum_{k=|T_{join}|-h+2}^{|T_{join}|-1} k)C_{me}$, where n comes from the last round, $h-1$ comes from the number of blinded keys that need to be calculated, and $|T_{join}|-h+1+i$

▷ T is the PFMH key tree of $\mathcal{G}$, and n is the size of T, T_{main} and T_{join} is the main tree and join tree of T.

if $((M \in T_{join})$ AND $(1 < |T_{join}| \leq log n))$ **then**
    SCENARIO I: Let P be M's sibling, remove M and M's parent from the key tree. If P has no children, change P's secret share, otherwise, change P's right child's secret share. Recursively update all the keys on P's key-path by applying multiple rounds of two-group DH.
else if $((M \in T_{join}$ AND $(|T_{join}| = 1$ OR $|T_{join}| > log n))$
        OR $(M \in T_{main}$ AND $|T_{join}| > 1)$
        OR $(M \in T_{main}$ AND M is the rightmost non-phantom leaf node)
        OR $(M \in T_{main}$ AND M has a phantom node in $T))$ **then**
    SCENARIO II: First, remove all phantom nodes and M from T. Second, apply the *split* procedure, and let
    $\mathcal{T} = \{T_1, \ldots, T_L\} = split(T)$. Third, change T_L's rightmost leaf node's secret share, and recursively update all the subgroup keys on this left node's key-path in T_L. Fourth, apply the unite procedure $unite(\mathcal{T})$.
else
    SCENARIO III: Find the rightmost non-phantom leaf node M' in T. Let P_{new} denote the node occupied by M, and P_{old} denote the node occupied by M'. M' moves to P_{new} and generates a new secret share for this location. If P_{old} lies in the join tree, then remove P_{old} and the root of T, otherwise, let P_{old} be M''s phantom node. Recursively update all the keys on P_{new}'s key path by applying multiple rounds of two-group DH.
end
All members update the key tree maintained by them locally according to the above key tree update procedure.

Algorithm 10: $Leave(\mathcal{G}, M)$

comes from the ith round. Since $|T_{join}| \leq \log n$, a loose upper-bound is $(n + \sum_{k=1}^{\log n} k)C_{me}$, or $(n + 0.5(\log n)^2)C_{me}$. Similarly, it is easy to check that the total communication cost in term of multicast is upper-bounded by $2(h-1)C_{multicast}$, and the total communication cost in term of unicast is upper-bounded by $(n + 0.5(\log n)^2)C_{unicast}$.

Fig. 4.13 shows one example of key tree update upon single user leave under this scenario. In this example user M_6 leaves the group where node $(1, 0)$ is the root of main tree and node $(1, 1)$ is the root of join tree. Since the size of join tree is 2, according to Procedure 10, the node representing M_6 will be directly removed from the key tree, M_5 changes its secret share, and a new group key will be generated by applying two-group DH between M_5 and the subgroup in the main tree.

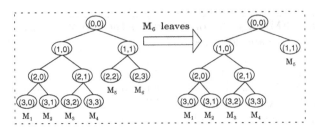

FIGURE 4.13. Example of key tree update upon single user leave under the first scenario

SCENARIO II: This scenario considers the case that any of the following situations happens:

1. The leaving member M is in the join tree, and the size of join tree is either larger than $\log n$ or equal to 1;

2. M is in the main tree, and the size of join tree is larger than 1;

3. M is in the main tree, and is the rightmost non-phantom leaf node;

4. M is in the main tree, and occupies a phantom node in the key tree.

In these situations, instead of removing M (as well as its phantom location) from the key tree and recursively updating all the keys on its key-path, the whole key tree will be reorganized to generate a new PF tree as the main tree, and the join tree is set to be empty. This will reduce the rekeying cost as well as maintain a good key tree structure. The basic procedure is to first remove all the phantom nodes in the existing key tree, then apply the *split* procedure to partition the remaining key tree into many small full key trees which are indexed according to their size and their locations in the original key tree. After changing a certain member's secret share, the *unite* procedure will be applied to combine these full key trees into a PF key tree. Finally, all members will update the key tree structure maintained by themselves according the above procedure.

It is worth noting that due to the special structure of the PFMH tree, the PFMH tree structure is maintained after removing some phantom nodes: According to Procedure 10 scenario III, only those leaf nodes on the right-most of the tree can be phantom nodes. In other words, all phantom nodes lie in the right-most part of the tree. It is easy to check that for any PF-tree, after removing any number of right-most leaf nodes and those corresponding non-leaf nodes, the remaining part is still a PF-tree.

Since all the remaining members (leaf nodes) know the exact structure of key tree, after applying the *split* procedure the set of obtained full key trees will be indexed in the same way by all group members. Since the total number of remaining members is less than n, according to Theorem 1 clause 1, the total time cost is upper-bounded by $\log n$. If situation 1, 2, or 3 happens, the total number of full key trees after applying the *split* procedure is upper-bounded by $\log(n) + |T_{join}|$. In this case, the total communication cost in term of multicast is upper-bounded by $2(\log(n) + |T_{join}|)C_{multicast}$. If situation 4 happens, the total communication cost in term of multicast is upper-bounded by $2(2\log n + |T_{join}|)C_{multicast}$, where the extra $2\log n C_{multicast}$ is due to the fact that the main tree can be split into at most $2\log n$ full trees.

Next we analyze the computation cost under this scenario, which is mainly incurred by the *unite* procedure. After applying the *split* procedure, for any size that is greater than 1, there exists no more than 1 full key tree with this size when situation 1, 2, or 3 happens, and there exists no more than 2 full key trees with this size when situation 4 happens. The *unite* procedure can be implemented in two steps. In the first step all the key trees with only one leaf node will first be combined together into a set of full key trees with different sizes. In the second step these full key trees will be combined together with the other full key trees obtained by applying the *split* procedure to get the final PF tree. We first consider the more probable case that $T_1 \geq n/2$ where T_1 is the largest full key tree obtained after applying the *split* procedure. According to Theorem 1 clause 3 and clause 5, in this case the computation cost is upper-bounded by $C_{me}(2.5n + 2(\log n + |T_{join}|) + |T_{join}|(\log(|T_{join}|) + 1))$, where the term $|T_{join}|(\log(|T_{join}| + 1))$ comes from merging the nodes from the join tree into a set of full key trees with different sizes. If $T_1 < n/2$, which is a less probable case, according to Theorem 1 clause 3 and clause 6, the total computation cost is upper-bounded by $(3n + 2(\log n + |T_{join}|) + |T_{join}|(\log(|T_{join}|) + 1))C_{me}$. Similarly, the total communication cost in term of unicast is upper-bounded by $(2.5n + |T_{join}|\log(|T_{join}|))C_{unicast}$ if $T_1 \geq n/2$ and is upper-bounded by $(3n + |T_{join}|\log(|T_{join}|))C_{unicast}$ if $T_1 < n/2$.

If condition 4 is satisfied, which is a very rare event, at most $(n + \log n)C_{me}$ extra computation cost is needed to first combine those full key trees with the same size into a set of larger full key trees and at most $nC_{unicast}$ extra communication cost in term of unicast is needed.

Fig. 4.14 shows four examples of key tree update upon single user leave under this scenario.

- The first example corresponds to situation 1: the leaving member M_6 is in the join tree, and the size of join tree with root $(1,1)$ is larger than $\log n$. In this example, after removing M_6 and applying the *split* procedure, 3 full key trees (subgroups) are obtained: $\{M_1, M_2, M_3, M_4\}$, $\{M_5\}$, $\{M_7\}$. The result of *unite* procedure has also been demonstrated.

- The second example corresponds to situation 2: the leaving member M_2 is in the main tree with root $(1,0)$, and the size of join tree with root $(1,1)$ is larger than 1. In this case after removing M_2 and applying *split*, three full key trees are obtained: $\{M_3, M_4\}$, $\{M_5, M_6\}$, $\{M_1\}$. The result of *unite* has also been illustrated in the right side of the figure.

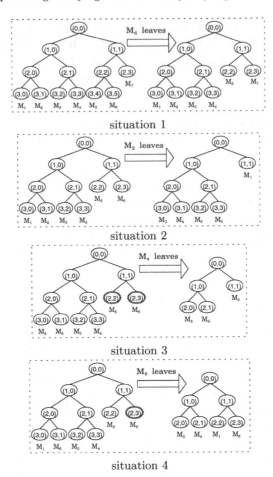

FIGURE 4.14. Examples of key tree update upon single user leave under the second scenario

- The third example corresponds to situation 3: the leaving member M_4 is in the main tree with root $(0,0)$ (the join tree is empty), and is the rightmost non-phantom leaf node, where nodes $(2,2)$ and $(2,3)$ are phantom nodes. In this case after removing the node representing M_4 and the phantom nodes and applying *split*, 2 full key trees are obtained: $\{M_5, M_6\}$ and $\{M_3\}$. The result of *unite* has also been illustrated in the right side of the figure.

- The fourth example corresponds to situation 4: the leaving member M_6 is in the main tree with root $(0,0)$ (the join tree is empty), and has occupied a phantom node $(2,3)$. In this case after removing the node representing M_4 and the phantom node and applying *split*, 3 full key trees are obtained: $\{M_3, M_4\}$, $\{M_1\}$ and $\{M_5\}$. The result of *unite* has also been illustrated in the right side of the figure.

TABLE 4.4. Rekeying cost bounds upon single user leave event

	communication cost $(C_{multicast})$	communication cost $(C_{unicast})$
Scenario 1	$O(2h - 2)$	$O(n + 0.5|T_{join}|^2)$
Scenario 2	$O(2 \log n + 2|T_{join}|)$	$O(2.5n + |T_{join}| \log(|T_{join}|))$
Scenario 3	$O(2 \log n)$	$O(n + 2|T_{left}|)$
	time cost (rounds)	computation cost (C_{me})
Scenario 1	$O(h)$	$O(n + 0.5|T_{join}|^2)$
Scenario 2	$O(\log n)$	$O(2.5n + |T_{join}| \log(|T_{join}|))$
Scenario 3	$O(\log n)$	$O(n + 2|T_{left}|)$

SCENARIO III: This scenario covers all the situations that neither of the first two scenarios can cover. Specifically, this scenario considers two situations: 1) M is in the main tree and the size of the join tree is 1; 2) the join tree is empty, and M is in the main tree and is not the right-most non-phantom node and does not have phantom node in the key tree. Under scenario III, the leaving member M is removed from the key tree, and M', which is the member who occupies the right-most non-phantom leaf node, moves to M's previous position, generates a secret share for this node, and recursively updates all the keys on this node's key-path. Now, M' occupies two positions, and the original position is called M''s phantom position. It is easy to check that the time cost is bounded by $\log n$, the communication cost in term of multicast is bounded by $2(\log n)C_{multicast}$, the computation cost is upper-bounded by $(n + 2|T_{left}| + \log n)C_{me}$, where T_{left} is T_{main}'s left subtree, and the total communication cost in term of unicast is upper-bounded by $(n + 2|T_{left}|)C_{unicast}$.

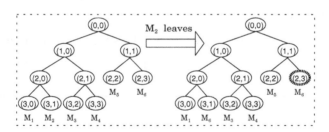

FIGURE 4.15. Example of key tree update upon single user leave under the third scenario

Fig. 4.15 shows one example of key tree update upon single user leave under this scenario. In this example the join tree is empty and the root of main tree is $(0,0)$. When user M_2 leaves the group, member M_6 will move to the location $(3,1)$ that previously represents M_2. Meanwhile, M_6 will also occupy node $(2,3)$ which now is a phantom node. M_6 will change its secret share and recursively update all the keys on its key-path, which is $\{(2,0), (1,0), (0,0)\}$.

Table 4.4 summarizes the rekeying cost upon single user leave events under different situations. Usually we have $|T_{left}| \geq n/2$, $h \simeq \frac{1}{2} \log n$,

$|T_{join}| \ll n$, and the average size of T_{left} is about $0.75n$. For the second and third scenario, in most cases the upper bound of computation cost can be simplified as $O(2.5nC_{me})$. For the first scenario, the bound of computation cost can be simplified as $O(nC_{me})$.

Group Merge and Group Partition Protocol

PACK also has group merge and group partition protocol to handle simultaneously join and leave of multiple users. Although multiple user events can be implemented by applying a sequence of single user join or leave protocols, such sequential implementations are usually not cost-efficient. Procedure 11 describes the group merge protocol, which combines two or more groups into a single group, and returns a PF key tree. Procedure 12 describes the group partition protocol, which removes multiple group member simultaneously from the current group and construct a new PF key tree for the rest of the group members.

▷ $T_1, \ldots, T_K$ are the key trees of $\mathcal{G}_1, \ldots, \mathcal{G}_K$;
Remove all *phantom* nodes from $T_1, \ldots, T_K$;
$T = unite(split(T_1) \bigcup \ldots \bigcup split(T_K))$;
Return T.

Algorithm 11: $merge(\{\mathcal{G}_1, \ldots, \mathcal{G}_K\})$

▷ T is the key tree of $\mathcal{G}$;
Remove all *phantom* members and members belonging to group $\mathcal{G}_1$ from T ;
$T = unite(split(T))$;
Return T.

Algorithm 12: $Partition(\mathcal{G}, \mathcal{G}_1)$

In the group merge protocols, after removing all phantom nodes from those key trees corresponding to different subgroups, each key tree is split into many full key trees. The final result is obtained by unite these full key trees into a PF tree following Procedure 7. Similar for the group partition protocol, after removing all phantom nodes and leaving nodes, the original key tree is split into many full key trees, and the unite procedure is then applied on these full key trees to create a PF key tree. Since the height of the returned tree is $\log n$, where n is the group size after merging/partitioning, the time cost of group merge/partition is bounded by $O(\log n)$. Obviously, the group merge and partition protocols have lower cost than the sequential implementations.

4.2.4   Performance Evaluation and Comparison

Forward and Backward Security

The group key secrecy means that attackers cannot obtain the group key even if they know all blind keys, which has been proved in the random-oracle model [74]. To show that PACK satisfies forward and backward secrecy, similar arguments as in [59] can be used, which have provided detailed proof for TGDH. PACK and TGDH use similar group key update procedures. The major difference between them are the underlying key tree structures which do not affect the security of the scheme. Therefore in this paper we will not provide detailed proof of forward and backward secrecy. Next we only roughly sketch the proof. We first consider backward secrecy. When a new user M wants to join the group, M picks its secret share r. After several rounds of two-group DH, M gets all blinded keys on its co-path, and it can compute all secret keys on its key-path using its own secret share and the blinded keys on its co-path. Clearly, all these keys contain M's secret share; hence they are independent of previous secret keys on that path. Therefore, M cannot derive any previous keys. The forward secrecy can be shown in a similar way. When a member M leaves the group, at least one current member changes its share, and all the keys on M's key path will be updated to remove M's secret share. Hence, M only knows at most all blinded keys, and the group key secrecy property prevents M from deriving any future group keys. By combining backward secrecy and forward secrecy, we can derive the key independence.

Cost Comparison

This section compares the rekeying cost in PACK upon single user join and leave events with two existing tree-based contributory group key agreement schemes: TGDH [59] and DST [60]. All three types of cost are considered: time, computation, and communication in term of multicast. Since in general members' leaving time is not known in advance, in DST, only join-tree is used. Table 4.5 lists the approximate bounds of different cost for the three schemes.

From the above comparison, we can see that PACK has the lowest cost in terms of time, computation, and communication. For example, for user join, only 1 or 2 rounds are needed in time cost, while DST needs $1 + \log \log n$ rounds and TGDH needs $\log n$ rounds. Similar results can also be seen in communication cost for user join. The total computation cost is computed as the average of user join cost and leave cost, DST has similar cost as TGDH, which is an order of $2n$, while for PACK, the order is from n to $1.75n$, with the saving ranging from 15% to 50% compared with DST and TGDH.

TABLE 4.5. Rekeying cost comparison among different schemes

	time cost	communication cost	computation cost
Upon Single User Join Event			
PACK	$1 \sim 2$	$2 \sim 4C_{multicast}$	nC_{me}
TGDH	$\log n$	$2(\log n)C_{multicast}$	$2nC_{me}$
DST	$1 + \log \log n$	$(1 + \log \log n)C_{multicast}$	$(n + \log n)C_{me}$
Upon Single User Leave Event			
PACK	$\log n$	$2(\log n)C_{multicast}$	$(1 \sim 2.5)nC_{me}$
TGDH	$\log n$	$2(\log n)C_{multicast}$	$2nC_{me}$
DST	$1 + \log n + \log \log n$	$2(1 + \log n + \log \log n) \cdot C_{multicast}$	$3nC_{me}$

Simulation Results

In the simulations, the user activities are generated according to the following probabilistic models: users join the group according to a Poisson process with average arrival rate λ, and users' staying time in the group follows an exponential distribution with mean μ (such a model is motivated by the user statistics in study of Mbone [72,73]). Then $\lambda\mu$ is the average number of users in the group, that is, the average group size. For each simulation, the initial group size is 0, λ is fixed, and μ varies to get different average group size configuration. For each configuration (different average group size), a sequence of $100\lambda\mu$ users join the group according to the Poisson process with rate λ, and each user's staying time is drawn independently from an exponential distribution with mean μ. In the simulations, the rekeying cost of the following three schemes: PACK, TGDH [66] and DST [60], are compared in all three aspects: computation, communication and time.

The simulation results are presented in Fig. 4.16. From these results we can see that upon single user join event, PACK has the lowest cost among all three schemes. Compared with DST, PACK has more than 10% reduction in computation cost, and more than 65% reduction in communication cost and time cost. Compared with TGDH, the reduction is even more, about 50% in computation cost and about 80% in time and communication cost. Upon single user leave event, compared with DST, PACK has about 25% reduction in computation cost, about 15% reduction in time cost, and has similar communication cost. Although PACK has slightly higher computation and communication cost than TGDH upon single user leave event, when averaged over both join and leave events, the reduction is still significant, with 20% reduction in computation cost, 35% reduction in communication cost, and 40% reduction in time cost.

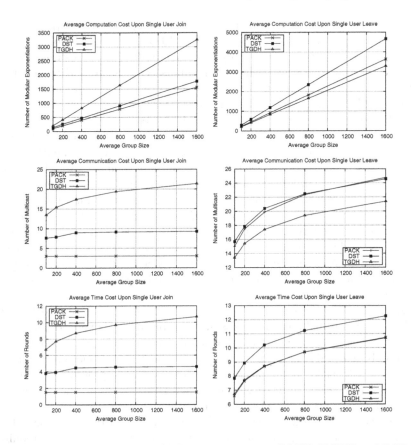

FIGURE 4.16. Comparison of rekeying cost among PACK, TGDH and DST

4.2.5 Contributory Group Key Agreement with Key Validation

In practice, there may exist malicious or compromised group members who do not perform key agreement protocol honestly and cause key generation failure. One example of key generation failure is group partition where some users share one key while the others share another different key. Therefore, besides the four security requirements discussed in Section 4.2.1, the group key management should also have the key validity property. That is, without being detected by other users, malicious users cannot prevent valid group key from being generated by providing false information. In this section, we discuss the possible damage that untruthful users can cause and the mechanisms to check the key validity.

When implementing the two group DH using the method described in Section 4.2.1, an untruthful member can cause key generation failure only if it has been elected as a delegate. In this case, an untruthful member, e.g., a in subgroup $\mathbf{A}$, can send false blinded key $f(K'_A)$ to selected members in subgroup $\mathbf{B}$. As a consequence, those members in $\mathbf{B}$ who have received false blinded keys from a cannot obtain the valid group key K_{AB}, that is, these members have been implicitly revoked from the new group.

We introduce two methods to check the validity of key establishment procedure and to detect malicious members. One is *preventive* and the other is *detective*. In the preventive scheme, for each group, m members are elected as delegates and broadcast the blinded key. Then each group member checks whether these m copies of blinded keys are same. Since all the keying messages have been signed by the senders, the member who has sent false information can be easily detected by other group member. In the detective scheme, after the each round of DH, m members are elected to broadcast a common known message encrypted using the newly generated group/subgroup key. Other members check whether they can use their new group/subgroup key to successfully decrypt the message. If a user cannot obtain this commonly known message after decryption, it broadcasts an error message that includes the blinded key and the messages it has received. Again, since keying messages are signed by their senders, those malicious members who have sent false blinded key or false encrypted messages can be detected.

Although colluders can compromise both preventive and detective schemes, the probability of successful collusion attack is very low because those m delegates or m users who broadcast the encrypted message are randomly selected. In addition, the detective method are more resistant to collusion attacks than the preventive methods. In the preventive method, the m delegates are selected within one subgroup, while in the preventive method, the m users are selected from both subgroups.

Key validation requires extra cost. In each round of two-group DH, the preventive scheme require $2m$ broadcast and the reactive scheme require m broadcast, m encryption and n decryption, where n is the size of the new subgroup after the DH round. It is noted that the extra cost due to checking is proportional to the cost of the key management schemes without the checking schemes. Thus, in previous analysis and comparisons, we did not count the extra cost associated with key validation.

4.3   Chapter Summary

In this chapter, we presented two contributory key agreement schemes: JET and PACK, which are both designed to achieve high efficiency in tree-based key agreement schemes.

JET reduces the latency in key agreement by utilizing a join-exit-tree structure, where the join and exit subtrees serve as temporary buffers for joining and leaving users. To achieve time efficiency, the optimal subtree capacity should be at the log scale of the group size. JET has an adaptive algorithm to activate and update join and exit subtrees. JET can achieve an average time cost of $O(\log(\log n))$ for user join and leave events in a group of n users, and reduces the total time cost of key update over a system's life time from $O(n \log n)$ by prior works to $O(n \log(\log n))$. JET also has low computation and communication overhead.

Inspired by JET, a better scheme called PACK was developed to further reduce overhead. PACK reduces the communication and computation overhead associated with key updating in two ways. First, it uses the novel PFMH tree structure that consist of a main tree, which is optimal for user leave, and a join tree, which is optimal for user join. Second, the concept of phantom user location in the PFMH allows the cost amortization when handling user leave. Upon single user join, PACK has the time cost as 1 or 2 rounds of two-group DH, the communication cost as 2 or 4 multicast, and the average computation cost as 1 modular exponentiation per user. Upon single user leave event, PACK takes at most $\log n$ rounds of two-group DH in terms of time cost, $O(\log n)$ multicast in communication cost, and an average of 2 modular exponentiations per user in computation cost, where n is the current group size. PACK achieves the performance lower bound derived in [62].

5

Optimizing Multicast Key Management for Cellular Multicasting

There has been significant advancements in building a global wireless infrastructure that will free users from the confines of static communication networks. Users will be able to access the Internet from anywhere at anytime. As wireless connections become ubiquitous, consumers will desire to have multicast applications running on their mobile devices. In order to meet such a demand, there has been increasing research efforts in the area of wireless multicast [75–77].

In wireless networks, where bandwidth is limited and transmission error rate is high, the design of key management schemes need to consider the transmission of the rekeying messages. When the design of key management schemes can take advantage of the broadcast nature of wireless media as well as the wireless network topology, the communication overhead introduced by key management can be reduced. As a direct consequence, the reliability of key distribution can be greatly improved.

In this chapter, some important properties of tree-based centralized key management scheme will be exploited. Based on these properties, the concept of topology-aware key management will be introduced. A specific design of such topology-aware key management scheme is then presented in detail, followed by performance evaluation.

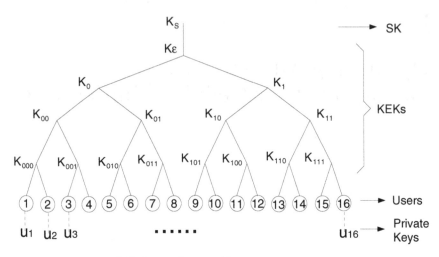

FIGURE 5.1. A typical key management tree

5.1 Targeting Property of Rekeying Messages

Let us revisit the tree-based centralized key management scheme described in Chapter 2. As a reference to the rest of the chapter, a key tree with 16 leaf node is drawn in Figure 5.1.

As described previously, many centralized key management schemes employ a tree hierarchy to maintain the keying material [8, 10, 78–80]. In the key tree example shown in Figure 5.1, each node of the key tree is associated with a key. The root of the key tree is associated with the session key (SK), K_s, which is used to encrypt the multicast content. Each leaf node is associated with a user's private key, u_i, which is only known by this user and the KDC. The intermediate nodes are associated with key-encrypted-keys (KEK), which are auxiliary keys and only for the purpose of protecting the session key and other KEKs. To make concise presentation, we do not distinguish the node and the key associated with this node in the remainder of the chapter.

In this example, user 16, that possesses $\{u_{16}, K_s, K_\epsilon, K_1, K_{11}, K_{111}\}$, leaves the group. Due to his departure, the KDC generates new keys and conveys new keys to the remaining users through a set of rekeying messages as:

- $\{K_{111}^{new}\}_{u_{15}}$: user 15 acquires K_{111}^{new},

- $\{K_{11}^{new}\}_{K_{111}^{new}}, \{K_{11}^{new}\}_{K_{110}^{old}}$: user 13,14,15 acquire K_{11}^{new},

- $\{K_1^{new}\}_{K_{11}^{new}}, \{K_1^{new}\}_{K_{10}^{old}}$: user 9, $\cdots$, 15 acquire K_1^{new},

- $\{K_\epsilon^{new}\}_{K_1^{new}}, \{K_\epsilon^{new}\}_{K_0^{old}}$: user 1, $\cdots$, 15 acquire K_ϵ^{new},

- $\{K_s^{new}\}_{K_\epsilon^{new}}$: all remaining users acquire K_s^{new}, where the notation x^{old} represents the old version of key x, x^{new} represents the new version of key x, and $\{y\}_x$ represents the key y encrypted by key x. This rekeying procedure is very similar to the procedure presented in [80].

It is seen that most rekeying messages are only useful to a subset of users, who are always neighbors on the key management tree. This property is referred to as the *targeting* property of the rekeying messages. In fact, the first rekeying message is only useful to user 15, the second rekeying message is only useful to users 13, 14, 15, the third rekeying message is useful to users $9, 10, \cdots, 15$, and the fourth and fifth rekeying messages are useful to all users.

It is noted that rekeying messages are usually delivered to all users through multicast communications. Because of the targeting property, rekeying messages do not have to be sent to every user in the multicast group.

5.2 Topology-aware Key Management

Due to the targeting property, the KDC can deliver the rekeying messages only to the users who need them. However, this does not necessarily reduce the communication overhead. The users who need the rekeying messages can randomly scatter in the network.

In order to take advantage of the targeting property, the design of the key management tree must consider the network topology. Particularly, the key tree need to match the network topology in such a way that the neighbors on the key tree are also physical neighbors on the network. As a consequence, when the KDC delivers the rekeying messages only to the users who need them, the delivery can be localized to a small regions of the network. Additionally, in order to achieve localized message delivery, it is necessary to have the assistance of entities that would control the rekeying message transmission.

Since the schemes that take advantage of the targeting property must consider the network topology in their design, these schemes are referred to as *topology-aware* key management schemes. The topology-aware key management schemes can localize the delivery of rekeying messages and therefore reduce the communication overhead associated with key management.

5.3 Topology-aware Key Management in Cellular Wireless Network

In this section, the design of a topology-aware key management scheme for cellular wireless network proposed in [81] is described in detail.

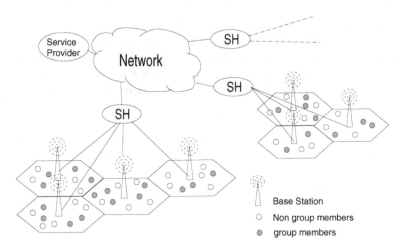

FIGURE 5.2. A cellular wireless network model

5.3.1 Key Tree Design

A cellular network model is depicted in Figure 5.2. This model, proposed in [82], consists of mobile users, base stations (BS) and supervisor hosts (SH). The SHs administrate the BSs and handle most of the routing and protocol details for mobile users. The service provider, the SHs, and the BSs are connected through high-speed wired connections, while the BSs and the mobile users are connected through wireless channels. In this work, the SHs can represent any entity that administers BSs, such as the region servers presented in [83] and radio network controllers (RNCs) in 3G networks [84].

In cellular wireless networks, multicast communication can be implemented efficiently by exploiting the inherent broadcasting nature of the wireless media [85–87]. In this case, multicast data is first routed to the BSs using multicast routing techniques designed for wireline networks [88], and then broadcast by the BSs to mobile users.

If both the SHs and the BSs can determine whether the rekeying messages are useful for the users under them, then the cellular wireless network has the capability of sending messages to a subset of users. In particular, the SHs multicast a rekeying message to their BSs if and only if the message is useful to one or several of their BSs, and the BSs broadcast the rekeying message to their users if and only if the message is useful to the users under them.

The information needed to identify whether a SH or BS needs a rekeying message can be sent in the rekeying message header. The size of this overhead information is typically small compared to the size of the actual rekeying messages.

The core of the topology-aware key management scheme is to design a key tree that matches the network topology. The design can be carried out in three steps.

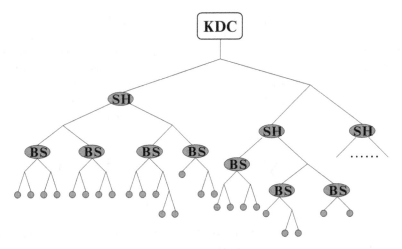

FIGURE 5.3. A Topology -matching key management tree

- Step 1: Design a subtree for the users under each BS. These subtrees are referred to as *user subtrees.*

- Step 2: Design subtrees that govern the key hierarchy between the BSs and the SH. These subtrees are referred to as *BS subtrees.*

- Step 3: Design a subtree that governs the key hierarchy between the SH and the KDC. This subtree is referred to as the *SH subtree.*

The combination of all subtrees is called the Topology-Matching Key Management (TMKM) tree. Figure 5.3 illustrates a TMKM tree for the network topology shown in Figure 5.2.

Classic key management trees, such as those in [8, 10, 79, 80], are independent of the network topology. These key trees can be referred to as the Topology Independent Key Management (TIKM) trees.

5.3.2 *Performance Metrics*

For performance evaluation, the communication burden caused by rekeying messages in the wired portion and in the wireless portion of the network should be studied separately.

Under each SH, the *wireline-message-size* is defined as the total size of the rekeying messages multicast by the SHs to the BSs, and the *wireless-message-size* is defined as the total size of the rekeying messages broadcast by the BSs. The message size is measured in units whose bit length is the same size as the key length. It is often assumed that the network connection between the KDC and the SHs has ample bandwidth resource and experience very low error rate. Thus, the wireline-message-size does not include the communication overhead between the KDC and the SHs.

Let S_1^l denote the wireline-message-size under the lth SH and S_2^l denote the wireless-message-size under the lth SH, where $l = 1, 2, \cdots, n_{sh}$ and n_{sh} is the total number of SHs. For example, when the length of the session key and KEKs is 128 bits each, if a 256 bit long rekeying message is multicast by the lth SH and then broadcast by 3 BSs under the lth SH, then $S_1^l = 2$ and $S_2^l = 6$. Assuming that users do not leave simultaneously, then the rekeying wireline cost, C_{wire}, the rekeying wireless cost $C_{wireless}$, and the total rekeying cost C_T, are defined as:

$$C_{wire} = \sum_{l=1}^{n_{sh}} \alpha_1^l E[S_1^l] \; ; \quad C_{wireless} = \sum_{l=1}^{n_{sh}} \alpha_2^l E[S_2^l]$$

$$C_T = \gamma \cdot C_{wireless} + (1 - \gamma) \cdot C_{wire} \tag{5.1}$$

where $E[.]$ indicates expectation over the statistics governing the user joining and leaving behavior. Here, $0 \le \gamma \le 1$ is the *wireless weight*, which represents the importance of considering the wireless cost, and $\{\alpha_1^l\}$ and $\{\alpha_2^l\}$ are the sets of weight factors that describe the importance of considering the wireline-message-size and wireless-messages-size under the lth SH respectively. When SHs administrate areas with similar physical network structure and channel conditions, $\{\alpha_1^l\}$ and $\{\alpha_2^l\}$ can be chosen as 1. In addition, the *combined-message-size* is defined as

$$S_T^l = \gamma \cdot S_2^l \alpha_2^l + (1 - \gamma) \cdot S_1^l \alpha_1^l. \tag{5.2}$$

Thus, C_T can also be expressed as $C_T = \sum_{l=1}^{n_{sh}} E[S_T^l]$.

For a given wireless weight γ, $\{\alpha_1^l\}$, and $\{\alpha_2^l\}$, both the TMKM and TIKM trees should be designed to minimize the total communication cost, C_T.

5.3.3 Handoff Schemes for TMKM Tree

In mobile environments, the user will subscribe to a multicast service under an initial host agent, and through the course of his service move to different cells and undergo *handoff* to different base stations. Although the user has moved, he still maintains his subscription to the multicast group. Since the TMKM tree depends on the network topology, the physical location of a user affects the user's position on the key management tree. When a user moves from one cell to another cell, the user needs to be relocated on the TMKM tree. In this section, the *handoff scheme* only refer to the process of relocating a user on the key tree.

One solution to the handoff problem is to treat the moving user as if he departs the service from the cell that he is leaving from and then rejoins the service in the cell that he has moved to. This scheme, referred to as the *simple handoff scheme*, is not practical for mobile networks with frequent handoffs since rekeying messages are sent whenever handoffs occur.

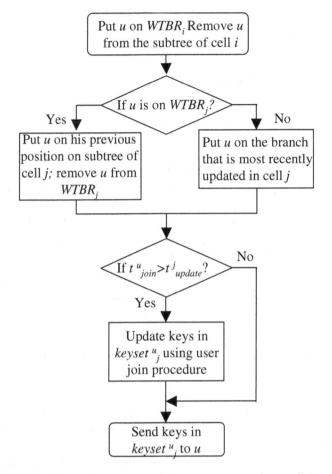

FIGURE 5.4. Key update process when user u moves from cell i to cell j

A more efficient handoff scheme defers the key update associated with handoffs until users' departure time. During handoff, if a user remains subscribed to the multicast group, it is not necessary to remove the user from the cell where he previously stayed. Allowing a mobile user to have more than one set of valid keys while he stays in the service does not compromise the requirements of access control, as long as all of the keys that he possesses are updated when he finally leaves the service.

In order to trace both the users' handoff behavior and the key updating process, a wait-to-be-removed (WTBR) list is employed for each cell. The WTBR list of the cell i, denoted by $WTBR_i$, contains the users who (1) possess a set of valid keys on the user subtree of cell i and (2) are currently in the service but not in cell i. These WTBR lists are maintained by the KDC.

Let t^i_{update} denote the time of the last key update that occurs due to a departure occurring in cell i, and let t^u_{join} denote the time when the user u first joins the service. In addition, $keyset^u_i$ is defined as the set of keys possessed by the user u while he is in cell i.

The *efficient handoff scheme* in [81] is illustrated in Figure 5.4 and Figure 5.5, as:

- When user u moves from cell i to cell j,

  1. Put u on the WTBR list of cell i, i.e. $WTBR_i$, and remove him from the user subtree of cell i.

  2. If u has been in cell j before and is on $WTBR_j$, put u back on the branch of the subtree that he previously belonged to and remove him from $WTBR_j$. If u is not on $WTBR_j$, put u on the most recently updated branch on the user subtree of cell j. We note that the set of keys associated with u's new position, $keyset^u_j$, was updated at time t^j_{update}.

  3. If $t^u_{join} > t^j_{update}$, the keys in $keyset^u_j$ are updated using the procedure for user join described in [80]. If $t^u_{join} \leq t^j_{update}$, the keys do not need to be updated.

  4. The keys in $keyset^u_j$ are sent to u through unicast.

  The purpose of step 3 is to prevent u from taking advantage of the handoff process to access the communication that occurred before he joined. To see this, let u join the service at $t^u_{join} = t_0$ in cell i, and then immediately move to cell j. After relocation, user u obtains keys in $keyset^u_j$ that is updated at time $t^j_{update} = t_0 - \Delta$, where Δ is a positive number. In this case, if we do not update the keys in $keyset^u_j$ and u has recorded the communication in cell j before joining, u will be able to decrypt the multicast content transmitted in $[t_0 - \Delta, t_0)$, during which time he is not a valid group member.

- When user u leaves the multicast service from cell j:

  1. The keys that are processed by u and still valid should be updated. In particular, the keys in $\{keyset^u_i, \forall i : u \in WTBR_i\}$ and $keyset^u_j$ are updated using the procedure for user departure in [80].

  2. Check other users on the WTBR lists that contain u. If u and another user u^* are both on $WTBR_i$, and $keyset^u_i = keyset^{u^*}_i$, remove u^* from $WTBR_i$. It is noted that u^* is removed from $WTBR_i$ when u^* does not have valid keys associated with cell i any more. Step 2 does not require extra rekeying messages.

  3. Remove u from all WTBR lists.

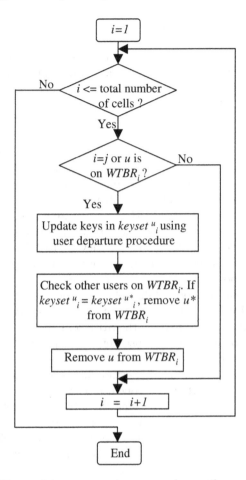

FIGURE 5.5. Key update process when user u leaves the service from cell j

A user will be removed from the WTBR lists not only when he leaves the service, but also when other users who share the same keys leave the service. Compared with the simple handoff scheme, the efficient handoff scheme can reduce the key updating caused by user relocation because the number of cells that need to update keys is smaller than the number of cells that a user has ever visited.

When the key tree matches with the network topology, handoffs result in users' relocation on the key tree, which inevitably introduce extra cost to the task of key management. Since the KDC often has significant computation and storage resources, there is no much concern about the cost for the KDC to maintain and update the WTBR lists. Instead, the major concern is the communication cost due to the fact that more than one set of keys may need to be updated for a departure user because of handoffs.

5.4   Performance Analysis

Matching the key management tree with the network topology has two contrasting effects on the rekeying message communication cost. First, the cost of sending one rekeying message is reduced because only a subset of the BSs broadcast the message. Second, the number of rekeying messages may increase due to handoffs. In this section, these two effects are analyzed.

To simplify the analysis, it is assumed that the system has a^{L_0} SHs, each SH administrates a^{L_1} BSs, and each BS has a^{L_2} users, where $a \geq 2$, L_0, L_1 and L_2 are positive integers. It is also assumed that the SHs administer areas with similar network structure and conditions. Therefore, $\{\alpha_1^l\}$ and $\{\alpha_2^l\}$ are approximated by 1. The user subtrees, BS subtrees, and SH subtree are designed as balanced trees with degree a and level L_2, L_1, and L_0, respectively. For fair comparison, the TIKM tree is also designed as an a-ary balanced tree with $(L_0 + L_1 + L_2)$ levels. In this chapter, the level of a tree is defined as the maximum number of nodes on the path from a leaf node to the root excluding the leaf node. Since the SHs are usually in charge of large areas, the probability of a user moving between SHs during a multicast service is much smaller than the probability of handoffs that are under one SH. In this analysis, the SH level handoffs is not considered. The communication cost is only caused by *one* departure user based on the rekeying procedure described in [8, 78, 80].

As illustrated by the example in Section 5.3, rekeying messages with size $(a \cdot L)$ need to be transmitted when one user leaves from a balanced key tree with degree a and level L. When using the TIKM tree, rekeying messages with size $a(L_0 + L_1 + L_2)$ are transmitted under a^{L_0} SHs and broadcast by $a^{L_0 + L_1}$ BSs. Therefore, when one user leaves the service, wireline-message-size, denoted by $\tilde{C}_w^{tikm}$, and the wireless-message-size, denoted by $\tilde{C}_{wl}^{tikm}$, are computed as

$$\tilde{C}_w^{tikm} = (aL_0 + aL_1 + aL_2)a^{L_0} \tag{5.3}$$
$$\tilde{C}_{wl}^{tikm} = (aL_0 + aL_1 + aL_2)a^{L_0 + L_1}. \tag{5.4}$$

The performance of the TMKM tree is affected by the user handoff behavior. Let random variable I denote the number of WTBR lists that contain the departing member when he leaves the service. The function $B(b, i, a)$ describes the number of intermediate KEKs that need to be updated. $B(b, i, a)$ is equivalent to the expected number of occupied boxes when putting i items in b boxes with repetition, where each box can have at most a items. A box is called occupied when one or more items are put into the box. The detailed calculation of $B(b, i, a)$ is described as follows.

Define $n(b, i, a)$ to be the number of non-empty boxes when randomly placing i identical items into b identical boxes with repetition, where each box can hold at most a items. Due to the definitions of $n(b, i, a)$ and $B(b, i, a)$, $B(b, i, a)$ is the expected value of $n(b, i, a)$, i.e. $B(b, i, a) = E[n(b, i, a)]$. It is obvious that $n(b, i, a)$ is bounded as $B_0 \leq n(b, i, a) \leq B_1$, where

$B_0 = \lceil \frac{i}{a} \rceil$ and $B_1 = \min(i, b)$. An intermediate quantity $w(y, i, a)$ is defined as the number of ways to put i items into y boxes such that each box contains at least 1 and at most a items. $w(y, i, a)$ can be calculated recursively as:

$$w(B_0, i, a) = \binom{aB_0}{i} \tag{5.5}$$

$$w(B_0 + k, i, a) = \binom{a(B_0 + k)}{i}$$

$$- \sum_{m=0}^{k-1} \binom{B_0 + k}{B_0 + m} w(B_0 + m, i, a), \tag{5.6}$$

where $0 \le k \le B_1 - B_0$. Then, the pmf of $n(b, i, a)$ can be expressed as:

$$Prob\{n(b, i, a) = B_0 + k\} = \frac{1}{N} \binom{b}{B_0 + k} w(B_0 + k, i, a), \tag{5.7}$$

where $N = \binom{ab}{i}$ represents the total number of ways of putting i items into b boxes. By substituting (5.6) into (5.7), one can see

$$Prob\{n(b, i, a) = B_0 + k\} = \frac{1}{N} \binom{b}{B_0 + k} \binom{a(B_0 + k)}{i}$$

$$- \sum_{m=0}^{k-1} \frac{\binom{b}{B_0+k}\binom{B_0+k}{B_0+m}}{\binom{b}{B_0+m}} Prob\{n(b, i, a) = B_0 + m\}.$$

It can be verified that:

$$\frac{\binom{b}{B_0+k}\binom{B_0+k}{B_0+m}}{\binom{b}{B_0+m}} = \binom{b - B_0 - m}{k - m}.$$

Therefore,

$$Prob\{n(b, i, a) = B_0 + k\} = \frac{1}{N} \binom{b}{B_0 + k} \binom{a(B_0 + k)}{i}$$

$$- \sum_{m=0}^{k-1} \binom{b - B_0 - m}{k - m} Prob\{n(b, i, a) = B_0 + m\}. \tag{5.8}$$

By substituting (5.5) into (5.7), one can get

$$Prob\{n(b, i, a) = B_0\} = \frac{1}{N} \binom{b}{B_0} \binom{aB_0}{i}. \tag{5.9}$$

Based on (5.8) and (5.9), one can calculate $Prob\{n(b, i, a) = B_0 + k\}$ for $k = 0, 1, \cdots, B_1 - B_0$ recursively. Then, $B(b, i, a)$ (i.e. $E[n(b, i, a)]$) is calculated as

$$B(b, i, a) = \sum_{k=0}^{B_1 - B_0} (B_0 + k) \cdot Prob\{n(b, i, a) = B_0 + k\}. \tag{5.10}$$

When one user leaves the service and he is on $I = i$ WTBR lists, one can verify that:

- $(i \cdot L_2)$ keys on user subtrees need to be updated. Thus, rekeying messages with total size $(iaL_2 - 1)$ are transmitted under one SH and broadcast by a single BS.

- $B(a^{L_1-m}, i, a^m)$ KEKs on the level $(L_1 - m)$ of the BS subtree need to be updated. Thus, messages with size $aB(a^{L_1-m}, i, a^m)$ are transmitted under one SH and broadcast by a^m BSs. Here, $m = 1, \cdots, L_1$, and the level 0 of a tree is just the root.

- (a^t) KEKs on the level $(L_0 - t)$ of the SH subtree need to be updated. Thus, messages with size (a^{t+1}) are sent under (a^t) SHs and broadcast by $(a^{L_1} \cdot a^t)$ BSs. Here, $t = 1, 2, \cdots, L_0$.

- In addition, one message is needed to update the session key K_s. This message is sent to all a^{L_0} SHs and $a^{L_0+L_1}$ BSs.

Therefore, when the departing user belongs to i WTBR lists, the expected value of the wireline-message-size, denoted by $C_w^{tmkm}(i)$, and the expected value of the wireless-message-size, denoted by $C_{wl}^{tmkm}(i)$, are computed as

$$C_w^{tmkm}(i) = iaL_2 + \sum_{m=1}^{L_1} aB(a^{L_1-m}, i, a^m) + \sum_{t=1}^{L_0} a^{t+1} \tag{5.11}$$

$$C_{wl}^{tmkm}(i) \;=\; iaL_2 - 1 + \sum_{m=1}^{L_1} a^{m+1} B(a^{L_1-m}, i, a^m)$$

$$+ a^{L_1} \sum_{t=1}^{L_0} a^{t+1} + a^{L_0+L_1}. \tag{5.12}$$

The performance of the TIKM tree and the TMKM tree can be compared by examining the values of $\tilde{C}_w^{tikm}$ and $C_w^{tmkm}(i)$, $\tilde{C}_{wl}^{tikm}$ and $C_{wl}^{tmkm}(i)$. In Figure 5.6, these values are plotted for different i and L_0, when the other parameters are fixed as $a = 2$, $L_1 = 3$, and $L_2 = 6$. Since the TIKM tree is not affected by handoffs, $\tilde{C}_w^{tikm}$ and $\tilde{C}_w^{tikm}$ are constant. Figure 5.6(a) and Figure 5.6(b) show the wireline-message-size and wireless-message-size respectively, when the system has only one SH. Figure 5.6(c) and Figure 5.6(d) show the corresponding curves for 2 SHs, while Figure 5.6(e) and Figure 5.6(f) depict the corresponding curves for systems with 8 SHs. It is observed that:

- Both $C_w^{tmkm}(i)$ and $C_{wl}^{tmkm}(i)$ are increasing functions of i.

- The TMKM tree always reduces the wireless-message-size, and this advantage becomes larger when the system contains more SHs.

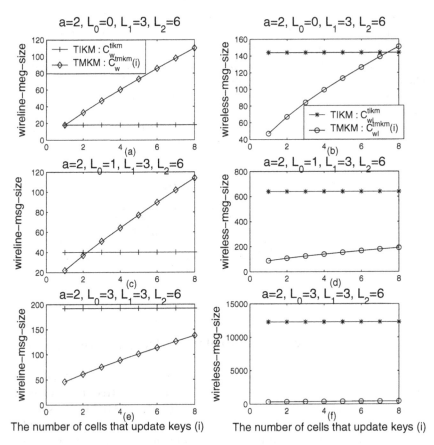

FIGURE 5.6. Comparison of the wireless cost and the wireline cost for one user departure

- For systems containing only one SH, i.e. $L_0 = 0$, the TMKM trees introduce larger wireline-message-size than TIKM trees due to the handoff effects. When there are multiple SHs, the TMKM scheme can take advantage of the fact that some SHs do not need to transmit rekeying messages to their BSs, and can reduce the wireline-message-size when i is small. It should be noted that the wireline cost will be larger than that given in (5.11) if there are SH-level handoffs.

Since TMKM trees reduce the wireless-message-size more effectively than reducing the wireline message size, a larger wireless weight γ leads to an improved advantage of TMKM trees over TIKM trees. Using large γ is a reasonable scenario since the wireless portion of the network usually experiences a higher error rate and has less available bandwidth when compared to the wireline portion, which makes the wireless cost the major concern in many realistic systems. In addition, the communication cost of the TMKM tree increases with the number of cells that need to update keys when a user leaves. Therefore, when handoffs are less likely to happen, the TMKM tree has larger advantage over the TIKM tree.

Scalability is another important performance measure of key management schemes [78]. Let $N = a^{L_0}$ denote the number of SHs. When $N \to \infty$, the scalability properties can be easily obtained from (5.3)-(5.12), and are summarized in Table 5.1. Both Figure 5.6 and Table 5.1 demonstrate that the communication cost of TMKM trees scales better than that of TIKM trees when more SHs participate in the multicast.

TABLE 5.1. Scalability comparison between TMKM and TIKM trees when the number of SHs(N)→ ∞.

	wireline-message-size	wireless-message-size
TIKM	$\sim aN \log_a N$	$\sim a^{L_1+1} N \log_a N$
TMKM	$\sim a^2 \log_a N$	$\sim a^{L_1+2} \log_a N$

5.5 Separability of the Optimization Problem

The TMKM tree consists of user-subtrees, BS-subtrees, and SH-subtrees. This could make the problem of finding optimal TMKM tree structure complicated. The good news is that optimizing the entire TMKM tree is equivalent to optimizing those subtrees individually. A proof will be provided in this section. This is desirable since optimizing the subtrees separately reduces the dimension of the search space for optimal tree parameters and significantly reduces the complexity of tree design.

In this proof, it is assumed that the users under the same SH have the same joining, departure and mobility behavior. Thus, the user subtrees under the same SH have the same structure. It is easy to verify that the main results in this section still hold in scenarios where the dynamic behavior of the users varies under different BSs.

In addition, it is assumed that the number of participating SHs and BSs do not change during the multicast service. In order to make the presentation more concise, the notation $D_{k,l}$ is introduced. $D_{k,l}$ represents the situation where k users are under the lth SH and one of these users leaves the service.

As discussed in Section 5.3, the total communication cost, C_T, is expressed as

$$C_T = \sum_{l=1}^{n_{sh}} E[S_T^l]. \tag{5.13}$$

Based on the definition of S_T^l, one can see that

$$E[S_T^l] = \sum_k p^l(k) G^l(k) E^l(k), \tag{5.14}$$

where

$p^l(k)$:   pmf of the number of users under the lth SH,

$G^l(k)$:   probability that a user leaves from the lth SH
given that k users are under the lth SH,

$E^l(k)$:   the expected value of the combined-message-
size given the condition $D_{k,l}$.

When a user leaves, the keys that need to be updated are divided into three categories: (1) the keys on the user subtrees, (2) the keys on the BS subtrees, and (3) the keys on the SH subtree. Under the condition $D_{k,l}$, let $A_1^l(k)$, $A_2^l(k)$ and A_3^l denote the expected value of the combined-message-size under the lth SH resulting from updating the keys on the user-subtrees, BS-subtrees and SH-subtrees, respectively. We note that A_3^l is not a function of k when there are no SH-level handoffs, and that $E^l(k) = A_1^l(k) + A_2^l(k) + A_3^l$. Then, (5.13) becomes

$$C_T = \sum_{l=1}^{n_{sh}} \left(\sum_k p^l(k) G^l(k) A_1^l(k) + \sum_k p^l(k) G^l(k) A_2^l(k) + A_3^l \cdot \left(\sum_k p^l(k) G^l(k) \right) \right).$$

The structure of the user-subtrees only affects $A_1^l(k)$, the structure of the BS-subtrees only affects $A_2^l(k)$, and the structure of the SH-subtrees only affects A_3^l. Therefore, for the TMKM tree, the user-subtrees, BS-subtrees and SH subtree can be designed and optimized separately. Particularly, the user-subtrees under the lth SH should be designed to minimize $\sum_k p^l(k) G^l(k) A_1^l(k)$, the BS subtree under the lth SH should be designed to minimize $\sum_k p^l(k) G^l(k) A_2^l(k)$, and the SH subtree should be designed to minimize $\sum_{l=1}^{n_{sh}} A_3^l \cdot \left(\sum_k p^l(k) G^l(k) \right)$.

5.6   Optimizing TMKM Tree Design

Key management schemes are closely related to the *key management architecture*, which describes the entities in the network that perform key management [78]. In cellular wireless networks, the BSs are not trusted to perform key management because they can be easily tampered with [82]. The SHs are able to perform key management if they are trusted and have the necessary computation and storage capabilities. The trustiness of the SHs depends on both the business model and the protection on the SHs. Based on whether SHs perform key management, the systems can be classified into two categories:

- In the first category, each SH performs key management for a subset of the group members who reside in the region where this SH is in charge. Each SH can be looked at as a local key distribution center. Without loss of generality, since the SHs are independent and may even adopt different key management schemes, one can study systems containing only one SH, which are referred to as *one-SH systems*.

- In the second category, SHs do not perform key management. Instead, there is a KDC that manages keys for all users. This KDC can be the service provider or a trusted third party. The systems containing many SHs are referred to as *multiple-SH systems*.

In one-SH systems, the TMKM tree consists of user-subtrees and a BS subtree. In multiple-SH systems, the TMKM tree consists of user-subtrees, BS-subtrees and a SH subtree.

5.6.1   Dynamic membership model

Before talking about optimizing the key tree design, one must specify application scenarios. For describing the application scenarios, a model is needed to describe the joining and leaving behavior of the users.

Mlisten [89] is a tool that can collect the join/leave times for multicast group members in MBone sessions. Using this tool, [90] [72] studied the characteristics of the membership dynamics of MBone multicast sessions and showed that the user arrival process can be modeled as Poisson and the membership duration of short sessions (that usually last several hours) is accurately modeled using an exponential distribution while the membership duration of long sessions (that usually last several days) is accurately modeled using the Zipf distribution [91]. Based on the population model of short MBone sessions, the following assumptions about the membership dynamics are made.

1. Under the lth SH, the user's arrival process is Poisson with rate λ_l and the service duration is governed by an exponential random variable with mean $1/\mu_l$, where $l = 1, 2, \cdots, n_{sh}$.

2. A user's joining and leaving behavior is independent of other users.

Based on the first assumption, the number of users under the lth SH is a Poisson random variable with rate θ_l, i.e. $p^l(k) = \frac{\theta_l^k}{k!}e^{-\theta_l}$, where $\theta_l = \lambda_l/\mu_l$ [92]. In addition, it can be shown that $G^l(k)$ approximately equals to $k \cdot \mu_l$. It is noted that the second assumption is reasonable in some types of multicast services, such as periodic news multicast, while it may not be correct for services such as a scheduled pay-per-view multicast, where different users are related with each other through watching the same content.

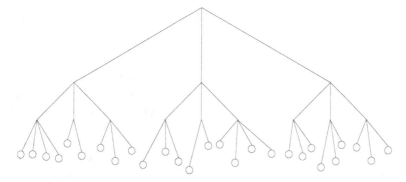

FIGURE 5.7. ALX tree

5.6.2   ALX tree structure

The TMKM scheme matches the key tree to the network topology by de-composing the key tree into user subtrees, BS subtrees, and SH subtrees. The TMKM scheme does not have constraints on the specific structure of these subtrees. Obviously, a balanced tree with per-determined degree is a valid choice for constructing subtrees. Another possible choice is a Huffman tree optimized for a given user statistics [61]. In both cases, major changes to the tree structure is expected when the group membership changes.

This section presents a new tree structure that is capable of handling membership additions, deletions, or relocations with minimal changes to the tree's structure. The advantages of the ALX tree will also be described.

As illustrated in Figure 5.7 and parameterized by the triple $(a, L, \mathbf{x})$, this $(a, L, \mathbf{x})$-logical tree has $L+1$ levels. The upper L levels, which comprise a full balanced subtree with degree a, are fixed during the multicast service. The users are represented by the leaf nodes on the $(L+1)^{st}$ level. A vector $\mathbf{x}$ is used to describe the $(L+1)^{st}$ level, where x_i is the number of users attached to the ith node of the Lth level, and $i = 1, 2, \cdots, a^L$. In the example shown in Figure 5.7, $\mathbf{x} = [4, 2, 3, 3, 2, 4, 3, 3, 3]$, $a = 3$ and $L = 3$. This tree structure is called as the ALX tree.

When using the ALX tree, the joining user is always put on the branch with the smallest value of x_i. The maximum number of users on an ALX tree is not restricted. When a user leaves, the average rekeying message size is $(\frac{k}{a^L} - 1 + aL)$, where k is the number of users on the ALX tree. When the user's arrival process is Poisson with rate λ, and the service time is an exponential random variable with mean $1/\mu$, the probability that a user leaves the key tree is approximately $k \cdot \mu$, and the pmf of k is $p(k) = \frac{\theta^k}{k!} e^{-\theta}$, where $\theta = \lambda/\mu$. The performance of the ALX tree is evaluated by the expected value of the rekeying message size, denoted by C_{alx}, and is calculated as:

$$C_{alx} = \sum_{k=1}^{\infty} p(k) \cdot k \cdot \mu \cdot (\frac{k}{a^L} - 1 + aL), \qquad (5.15)$$

It follows that the optimization problem of the ALX tree can be formulated as:

$$\tilde{C}_{alx} = \min_{a>1,L>0} C_{alx}. \tag{5.16}$$

Balanced trees whose degree is pre-determined, such as binary and trinary trees, are widely used to design key trees [78,80]. These trees are referred to as the fixed-degree trees. The ALX tree is compared with the fixed-degree trees as follows.

Adding or removing a user from balanced fixed-degree trees often requires splitting or merging nodes. For example, when a new user is added to the key tree shown in Figure 5.1, one leaf node must be split to accommodate the joining user. In this case, a new KEK is created and must be transmitted to at least one existing user. When using the ALX tree structure, however, no new KEKs are created during membership changes. Updating existing KEKs for user join can be achieved without sending any rekeying messages, as suggested in [80], because existing users can update KEKs using one-way functions after being informed of the need to update their keys. Therefore, the ALX tree structure allows for a key updating operation that does not require sending any rekeying messages during user joins. In addition, the ALX tree introduces minimal change to the tree structure with dynamic membership and therefore is easy to implement and analyze.

On the other hand, the ALX tree is optimized over the distribution of the group size. If we takes individual snapshots of the system when the group size is very small or large, the ALX tree may not perform as well as fixed degree trees that adjust themselves according to the group size. However, the cost of ALX trees, denoted by $\tilde{C}_{alx}$, is in fact very close to the performance lower bound of fixed degree trees.

Similar to (5.15), the expected rekeying message size when using a tree with fixed degree n, denoted by $C_{fix}(n)$, is calculated as:

$$C_{fix}(n) = \sum_{k=1}^{\infty} p(k) \cdot k \cdot \mu(n-1+n \cdot (P-1)) \,,$$

where P is the average length of branches for a tree with k leaves and degree n. It is well known that P equals the expected codeword length of a source code containing k symbols with equal probability. The bounds on P are known to be $\log_n(k) \leq P < \log_n(k) + 1$ [30]. Therefore,

$$C_{fix}(n) > \sum_{k=1}^{\infty} p(k) \cdot k \cdot \mu \cdot (n\log_n(k) - 1). \tag{5.17}$$

Based on (5.17), the performance lower bound for the fixed degree trees is given by

$$\tilde{C}_{fix} = \min_n \sum_{k=1}^{\infty} p(k) \cdot k \cdot \mu \cdot (n\log_n(k) - 1). \tag{5.18}$$

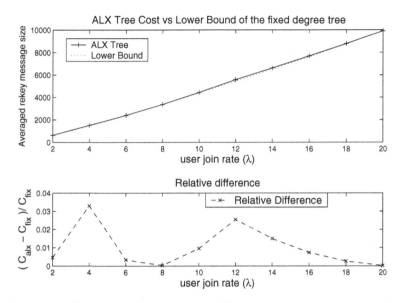

FIGURE 5.8. Comparison between the ALX tree performance and the lower bound for different user joining rates

It is noted that no fixed degree trees can reach this lower bound. In fact, $\tilde{C}_{fix}$ would be achieved if and only if one could (1) reorganize the tree immediately after user join or departure in such a way that the rekeying message size for the next user join/leave operation is minimized; and (2) reorganize the tree without adding any extra communication cost. However, reorganizing trees, such as splitting or merging nodes, requires sending extra keying information to users. These above two conditions can never be achieved simultaneously.

The lower bound in (5.18) is used as a reference for evaluating the performance of the ALX tree. In Figure 5.8, $\tilde{C}_{fix}$ and $\tilde{C}_{alx}$ are compared for different user joining rates, λ. In Figure 5.9, $\tilde{C}_{fix}$ and $\tilde{C}_{alx}$ are compared for different average service duration, $1/\mu$. It is observed that the relative difference between the lower bound and the performance of the ALX tree is less than 3.5%.

The ALX tree has the advantage of maintaining tree structure as user join and leaves, while its performance is very close to the lower bound of fixed degree trees. Although the ALX tree is not the optimal solution amongst all possible tree structures, its practical nature makes the ALX tree an ideal candidate for designing the user and BS subtrees.

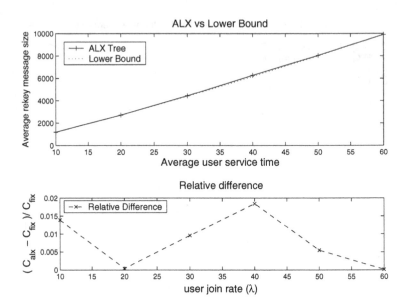

FIGURE 5.9. Comparison between the ALX tree performance and the lower bound for different average service duration

5.6.3   User subtree design

The user subtrees can be designed as ALX trees. Under the lth SH, the optimal tree parameters, a and L, solve

$$\min_{a,L} \sum_{k} p^l(k) G^l(k) A_1^l(k), \qquad (5.19)$$

where a and L are positive integers and $G^l(k) \approx k\mu_l$. Let $T_w^u(k,i)$ and $T_{wl}^u(k,i)$ respectively represent the expected value of the wireline-message-size and wireless-message-size caused by updating keys on the user subtrees, given that k users are under the lth SH, one of them leaves and he is on i WTBR lists. One can calculate

$$T_w^u(k,i) = T_{wl}^u(k,i) = (\frac{k/n_{bs}^l}{a^L} - 1 + aL)i.$$

Then, $A_1^l(k)$ is computed as

$$
\begin{aligned}
A_1^l(k) &= \sum_{i=1}^{n_{bs}^l} p_h^l(i)(\alpha_2^l \gamma T_{wl}^u(k,i) + \alpha_1^l(1-\gamma)T_w^u(k,i)) \\
&= (\alpha_2^l \gamma + \alpha_1^l(1-\gamma))(\frac{k/n_{bs}^l}{a^L} - 1 + aL)E[I^l], \qquad (5.20)
\end{aligned}
$$

where $E[I^l] = \sum_{i=1}^{n_{bs}^l} p_h^l(i)i$, and α_1^l and α_2^l are defined in Section 5.3. By substituting (5.20) into (5.19), the optimization problem for the user-subtrees under the lth SH is

$$\min_{a,L} \sum_k k \cdot p^l(k) \cdot (\frac{k/n_{bs}^l}{a^L} - 1 + aL). \tag{5.21}$$

The optimum a and L can be obtained by searching the space of possible a and L values.

5.6.4 BS subtree design

The BS subtrees can also be designed as ALX trees. The degree and the level of a BS subtree are denoted by a_{bs} and L_{bs}, respectively. Let $T_w^b(k, i)$ and $T_{wl}^b(k, i)$ respectively denote the expected value of the wireline-message-size and wireless-message-size caused by key updating on the BS subtree under the lth SH given the condition $D_{k,l}$ and the condition that the departing member is on i WTBR lists. One can calculate:

$$T_w^b(k, i) = s \cdot B(a_{bs}{}^{L_{bs}}, i, s) + \sum_{m=1}^{L_{bs}} a_{bs} \cdot B(a_{bs}{}^{L_{bs}-m}, i, s \cdot a_{bs}^m) \tag{5.22}$$

$$T_{wl}^b(k, i) \approx s^2 \cdot B(a_{bs}{}^{L_{bs}}, i, s)$$
$$+ \sum_{m=1}^{L_{bs}} a_{bs} \cdot a_{bs}{}^m \cdot s \cdot B(a_{bs}{}^{L_{bs}-m}, i, s \cdot a_{bs}^m), \tag{5.23}$$

where $s = \frac{n_{bs}^l}{a_{bs}^{L_{bs}}}$. Equation (5.22) and (5.23) are derived based on the following intermediate results:

- On average, $B(a_{bs}{}^{L_{bs}-m}, i, s \cdot a_{bs}^m)$ keys need to be updated on level $(L_{bs} - m)$ of the BS subtree.

- To update one KEK at level L_{bs}, the average message size is (s) and these messages are broadcast to an average of (s) BSs. To update one KEK at level $(L_{bs} - m), m > 0$, the message size is (a_{bs}) and these messages are broadcast by (a_{bs}^m) BSs.

From the definition of A_2^l and using both (5.22) and (5.23), we can see that

$$A_2^l = \sum_{i=1}^{n_{bs}^l} p_h^l(i)(\alpha_2^l \gamma T_{wl}^b(k, i) + \alpha_1^l(1 - \gamma)T_w^b(k, i))$$
$$= \sum_{i=1}^{n_{bs}^l} p_h^l(i) \Bigg(B(a_{bs}{}^{L_{bs}}, i, s) \left(s^2 \alpha_2^l \gamma + s\alpha_1^l(1 - \gamma) \right) \tag{5.24}$$
$$+ \sum_{m=1}^{L_{bs}} B(a_{bs}{}^{L_{bs}-m}, i, sa_{bs}^m)a_{bs} \left(a_{bs}^m s\alpha_2^l \gamma + \alpha_1^l(1 - \gamma) \right) \Bigg),$$

where n_{bs}^l is the number of BSs under the lth SH. In practice, it is difficult to obtain an analytic expression for $p_h^l(i)$ that depends on the statistical behavior of the users during membership joins and departures, as well as their mobility behavior and how handoffs are addressed.

Let random variable $\tilde{I}^l$ denote the number of cells that a leaving user has ever visited. Obviously, $\tilde{I}^l \geq I^l$. The pmf of $\tilde{I}^l$, denoted by $\tilde{p}_h^l(i)$, can be derived from user mobility behavior and the distribution of the service duration. For example, let t_M denote the service duration, t_n denote the new cell dwell time, and t_h denote the previously handed-off cell dwell time [93]. When the movement of the users follows the mobility model described in Section 5.7, the distributions of t_n and t_h are derived in [93]. The distribution of t_M is often assumed to follow an exponential distribution. Given these distributions, one can calculates $p_n = Prob\{t_M < t_n\}$ and $p_h = Prob\{t_M < t_h\}$. The number of cells that a user ever visited before departure, denoted by $\tilde{I}$, has the pmf as $Prob\{\tilde{I} = 1\} = p_n$, $Prob\{\tilde{I} = 2\} = (1 - p_n)p_h$, $Prob\{\tilde{I} = 3\} = (1 - p_n)(1 - p_h)p_h$, and $Prob\{\tilde{I} = i\} = (1 - p_n)(1 - p_h)^{i-2}p_h$.

Let $\tilde{A}_2^l$ denote the right hand side value in (5.24) when replacing $p_h^l(i)$ by $\tilde{p}_h^l(i)$. It can be verified that $\tilde{A}_2^l$ is an upper bound of A_2^l. This upper bound, $\tilde{A}_2^l$, is not a function of k.

As discussed in Section 5.5, the parameters of the BS subtree under the lth SH should be chosen to minimize $\sum_k p^l(k)G^l(k)A_2^l$. Since $G^l(k)$ is not a function of a_{bs} and L_{bs}, minimizing $\sum_k p^l(k)G^l(k)A_2^l$ is equivalent to minimizing A_2^l. Due to the unavailability of $p_h^l(i)$, the optimization problem of the BS subtree is formulated as

$$\min_{a_{bs}>1, L_{bs}>0} \tilde{A}_2^l. \tag{5.25}$$

5.6.5   SH subtree design

In a typical cellular network, each SH administrates a large area where both the user dynamics and the network conditions may differ significantly from the areas administered by other SHs. The heterogeneity among the SHs should be considered in designing the SH subtree. Due to SH heterogeneity, the ALX tree structure, which treats every leaf equally, is not an appropriate tree structure to build the SH subtree. Instead, the SH heterogeneity may be addressed by building a tree where the SHs have varying path lengths from the root to their leaf node.

The root of the SH subtree is the KDC, and the leaves are the SHs. The design goal is to minimize the third term in (5.15), which shall be denoted by C_{sh} and is

$$C_{sh} = \sum_{l=1}^{n_{sh}} q_l \cdot A_3^l, \tag{5.26}$$

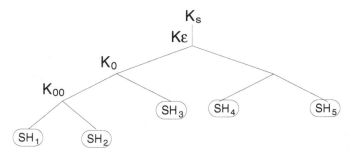

FIGURE 5.10. An example of the SH subtree

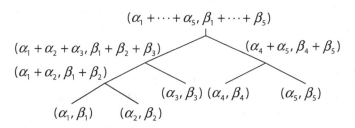

FIGURE 5.11. The cost pairs on the SH subtree

where $q_l = \sum_k p^l(k)G^l(k)$. Let β_l denote the communication cost of transmitting one rekeying message to all the users under the lth SH. Based on the definition of α_1^l and α_2^l in Section 5.3, it is easy to show that $\beta_l = (1-\gamma)\alpha_1^l + \gamma n_{bs}^l \alpha_2^l$.

The value of A_3^l can be calculated directly from β_l where $l = 1, 2, \cdots, n_{sh}$. In the simple example demonstrated in Figure 5.10, when a user under SH_1 leaves the multicast service, K_{00}, K_0, K_ϵ and K_s, need to be updated. The communication cost of updating K_{00} is $2(\beta_1 + \beta_2)$. The communication cost of updating K_0 is $2(\beta_1 + \beta_2 + \beta_3)$. The communication cost of updating K_ϵ is $2(\beta_1 + \beta_2 + \beta_3 + \beta_4 + \beta_5)$. Since the communication cost of updating K_s does not depend on SH subtree structure, it is not counted in the total communication cost. Then,

$$A_3^1 = 2(\beta_1 + \beta_2) + 2(\beta_1 + \beta_2 + \beta_3) + 2(\beta_1 + \beta_2 + \beta_3 + \beta_4 + \beta_5).$$

The goal of the SH subtree design is to find a tree structure that minimizes C_{sh} given β_l and q_l. However, it is very difficult to do so based on (5.26). Thus, C_{sh} is computed in a different way.

Let the SH subtree have the fixed degree n. A *cost pair*, which is a pair of positive numbers, can be assigned to each node on the tree. The cost pair of the leaf node that represents the lth SH is (q_l, β_l). The cost pair of the intermediate nodes are the element-wise summation of their children nodes' cost pairs, as illustrated in Figure 5.11. The cost pairs of all intermediate nodes are represented by (x_m, y_m), where $m = 1, 2, \cdots, M$, and M is the total number of intermediate nodes on the tree. Then, C_{sh} can be calculated as

$$C_{sh} = n \sum_{m=1}^{M} x_m \cdot y_m. \tag{5.27}$$

It is easy to verify that (5.27) is equivalent to (5.26). Based on (5.27), a tree construction method for $n = 2$ is developed.

1. Label all the leaf nodes using their cost pairs, and mark them to be active nodes.

2. Choose two active nodes, (x_i, y_i) and (x_j, y_j), such that $(x_i + x_j) \cdot (y_i + y_j)$ is minimized among all possible pairs of active nodes. Mark those two nodes to be inactive and merge them to generate a new active node with the cost pair $(x_i + x_j, y_i + y_j)$.

3. Repeat step 2 until there is only one active node left.

This method, which we call the greedy-SH subtree-design (GSHD) algorithm, can be easily extended to $n > 2$ cases. The GSHD algorithm produces the optimal solution when $\beta_1 = \beta_2 = \cdots \beta_{n_{sh}}$. However, it is not optimal in general cases. Since the optimization problem for the SH-subtree is non-linear, combinatorial, and even does not have a closed expression for the objective function, it is not computationally partial to seek the optimal SH subtree structure. Section 5.7 will present a comparison between the GSHD algorithm and the optimal solution obtained by exhaustive search.

5.7 Performance Evaluation

5.7.1 One-SH systems

This section provides a performance comparison between the TMKM tree and the TIKM tree in one-SH systems through both analysis and simulations.

The simulation setup is as follows. The network topology is a homogeneous cellular network that consists of 12 concatenated cells. The cell pattern is wrapped to avoid edge effects. The mobility model is set according to [93]. R denotes the radius of the cells, and V_{max} denotes the maximum speed of the mobile users. Since the wireless connection usually experiences a high transmission error rate and the number of users under one BS is larger than the number of BSs, the wireless communication cost of the multicast communication is assigned a larger weight than the wireline communication cost, i.e. $\gamma > 0.5$.

For the purpose of fair comparison, the TIKM tree is designed as an ALX tree, which is optimized for the statistics of the number of participating users. The wireline cost of the TIKM tree, denoted by C_{wire}^{tikm}, is computed using (5.15), where $p(k)$ denotes the pmf of the number of users

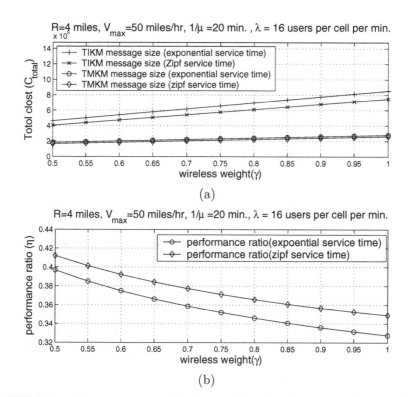

FIGURE 5.12. (a) The total message size as a function of the wireless weight; (b) Performance ratio as a function of the wireless weight

in the multicast service. The wireless cost of the TIKM tree is computed as $C_{wireless}^{tikm} = n_{bs} C_{wire}^{tikm}$, where n_{bs} is the total number of BSs. In one-SH systems, the total communication cost is $C_T^{tikm} = \gamma C_{wireless}^{tikm} + (1-\gamma) C_{wire}^{tikm}$.

The *performance ratio* η is defined as the total communication cost of the TMKM tree divided by the total communication cost of the TIKM tree, i.e. $\eta = C_T^{tmkm} / C_T^{tikm}$. When η is less than 1, the TMKM tree has smaller communication cost than the TIKM tree, and smaller η indicates an improved advantage that the TMKM tree has over the TIKM tree.

Figure 5.12(a) shows the total communication cost of the TMKM tree and the TIKM tree for different wireless weights (γ), when the cellular cells have a radius of 4 miles, the maximum mobile speed is 50 miles/hour, and the user joining rate is 16 users per minute per cell. The corresponding performance ratio is shown in Figure 5.12(b). In this simulation, two models are used to describe users' join/departure behavior. The first one, representing short sessions, uses a Poisson arrival and exponential service time duration model. The second one, representing long sessions, uses a Poisson arrival and Zipf service time duration model. The users stay in the service for an average of 20 minutes in both cases. Three observations are made.

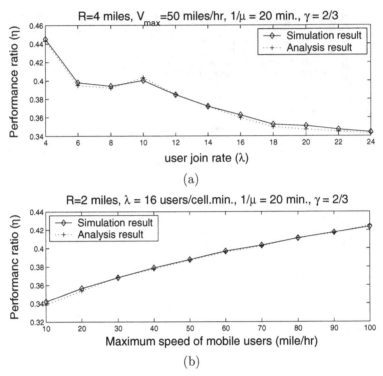

FIGURE 5.13. (a) Performance ratio for different user join rate; (b) Performance ratio for different users' maximum speed.

First, the communication cost of the TMKM tree is always less than 42% of the communication cost of the TIKM tree. Second, the performance ratio η is smaller for larger γ, which supports the argument in Section 5.4 that the advantage of the TMKM tree is larger when more emphasis is placed on the wireless cost. Third, when the wireless transmission is the bottleneck of the system, i.e. $\gamma = 1$, the TMKM tree can reduce the communication burden by as much as 65%, i.e. $\eta = 35\%$. In addition, two models yield similar results, which indicates that the performance of the TMKM is not sensitive to the models. For the remainder of the experiements, the short session model is adopted.

Figure 5.13(a) shows both the analysis and the simulation results of η for different user join rates (λ) when the radius of the cellular cells is 4 miles, the maximum mobile speed is 50 miles per hour, the average service time ($1/\mu$) is 20 minutes, and $\gamma = 2/3$. Since the exact expression for the pmf of I^l is not available, for the calculation of the analytical results, an empirically estimated pmf of I^l is obtained from simulations with the same user join/departure and mobility models. Figure 5.13 shows that the advantage of the TMKM tree is larger when the system contains more users. This property can be verified by the cost functions derived in the previous

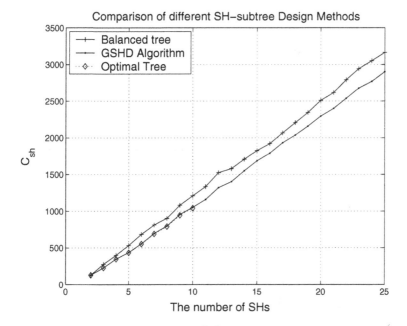

FIGURE 5.14. Comparison among SH subtree design methods

sections. In Figure 5.13(b), the performance ratio is shown for different V_{max} when the user joining rate is 16 users per minute per cell. The performance ratio is an increasing function of V_{max} when other parameters are fixed since handoffs occur more frequently as users move faster.

5.7.2   SH subtree design methods

This section provides a comparison among the GSHD algorithm, the optimal tree, and an alternative design. The optimal tree is obtained through exhaustive search. The alternative design is a balanced tree that treats each SHs equally just as in traditional key management schemes [].

In the simulation, half of the $\{\beta_l\}$ are uniformly distributed between 1 and 20, which represent rural areas, and the other half of $\{\beta_l\}$ uniformly distributed between 101 and 120, which represent metropolitan areas. Additionally, q_l is assumed to be proportional to β_l, and $\{q_l\}$ are normalized such that $\sum q_l = 1$, where $l = 1, 2, \cdots, n_{sh}$. (q_l is defined in 5.6.5 and represents the probability of a user leaving.)

In Figure 5.14, the communication cost caused by updating keys on SH-subtrees, C_{sh}, is shown when using different SH subtrees. Results are averaged over 500 realizations. Since exhaustive search is very computationally expensive, it is only done for 10 and fewer SHs. The simulation results indicate that the performance of the GSHD is very close to optimal. Compared

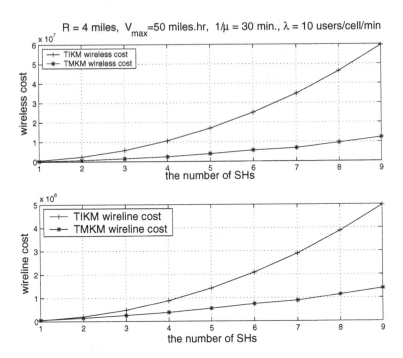

FIGURE 5.15. Performance comparison in multiple-SH systems with identical SHs

with the balanced tree, GSHD algorithm reduces the communication cost contributed by the SH subtree by up to 18%.

5.7.3 Multiple-SH systems

When the system contains multiple SHs that do not perform key management, the design of the TMKM tree should consider the topology of the SHs.

In the simulations, user-subtrees and BS-subtrees are designed as ALX trees and the SH-subtree a binary tree constructed using the GSHD algorithm. Each SH administers 12 concatenated identical cells. The simulation is performed for two cases.

In case 1, the user statistics and network conditions are identical under all SHs. Thus, α_1^l's and α_2^l's are set to be 1. The radius of the cells is $R = 4$ miles, the maximum velocity is $V_{max} = 50$ miles/hr, $\mu_l = 1/30$, and $\lambda_l = 10$ for all SHs.

In Figure 5.15, the wireless cost and the wireline cost of the TMKM trees and the TIKM tree are shown for different quantities of participating SHs. It is seen that the TMKM trees have both smaller wireless cost and smaller wireline costs than the TIKM trees when the number of SHs are equal or greater than 2, and the advantages of the TMKM trees are more significant

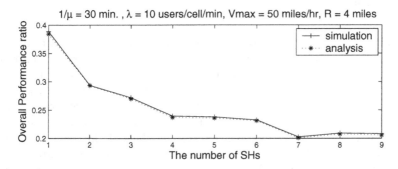

FIGURE 5.16. Performance comparison in multiple-SH systems with non-identical SHs

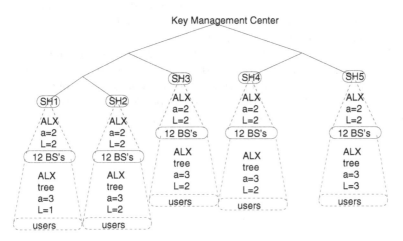

FIGURE 5.17. A TMKM tree containing 5 SHs

when the system contains more SHs, which verifies the analysis in Section 5.4. In addition, the corresponding performance ratio is drawn in Figure 5.16 for $\gamma = 2/3$. In this system, the communication cost of the TMKM trees can be as low as 20% of the communication cost of the TIKM trees. This indicates an 80% reduction in the communication cost.

In case 2, a more complicated system containing 5 SHs with different user joining rates is simulated. In this scenario, the λ_l values for the five SHs were set to 5, 10, 15, 20 and 25 respectively, and $R = 4$ miles, $V_{max} = 50$ miles/hr, and $\mu_l = 1/20$ for all SHs. The TMKM tree structure is shown in Figure 5.17. The TIKM tree is simply an ALX tree with degree 3 and level 6. In this system, the wireless cost of the TMKM tree is 21.8% of that of the TIKM tree, and the wireline cost of the TMKM tree is 34.0% of that of the TIKM tree. When the wireless weight γ is set to 2/3, the TMKM tree reduced total communication cost by 74%.

5.8   Chapter Summary

In this chapter, we described a topology-aware multicast key management scheme for mobile wireless environment. Compared with traditional tree-based key management schemes that are independent of network topology, the proposed TMKM scheme achieved a significant reduction in the communication burden associated with rekeying. The proposed key tree consists of user-subtrees, BS-subtrees and SH-subtrees. We proved that the problem of optimizing the communication cost for the TMKM tree is separable and can be solved by optimizing each of those subtrees separately. This property greatly reduced the complexity in key tree design. The ALX tree structure, which easily adapts to changes in the number of users, was introduced to build user-subtrees and BS-subtrees. The GSHD algorithm, which considers the network heterogeneity where the SHs administer areas with varying network conditions, was introduced to build the SH subtree. An efficient handoff scheme was introduced to address the consequences that user mobility has upon the TMKM tree. Both simulations and analysis demonstrated that the proposed TMKM scheme can significantly reduces the communication cost. In addition, the communication cost of the TMKM tree scales better than that of topology-independent trees as the number of participating SHs increases.

6

Key Management and Distribution for Securing Multimedia Multicasts

The distribution of identical data to multiple parties using the conventional point-to-point communication paradigm makes inefficient usage of resources. The redundancy in the copies of the data can be exploited in multicast communication by forming a group consisting of users who receive similar data, and sending a single message to all group users [1]. Access control to multicast communications is typically provided by encrypting the data using a key that is shared by all legitimate group members. The shared key, known as the session key (SK), will change with time, depending on the dynamics of group membership as well as the desired level of data protection. Since the key must change, the challenge is in key management–the issues related to the administration and distribution of keying material to multicast group members.

In order to update the session key, a party responsible for distributing the keys, called the group center (GC), must securely communicate with the users to distribute new key material. The GC shares keys, known as key encrypting keys (KEKs), that are used solely for the purpose of updating the session key and other KEKs with group members.

As an example of key management, we present a basic example of a multicast key distribution scheme. Suppose that the multicast group consists of n users and that the group center shares a key encrypting key with each user. Upon a member departure, the previous session key is compromised and a new session key must be given to the remaining group members. The GC encrypts the new session key with each user's key encrypting key and sends the result to that user. Thus, there are $n - 1$ encryptions that must be performed, and $n - 1$ messages that must be sent on the network.

The storage requirement for each user is 2 keys while the GC must store $n + 1$ keys. This approach to key distribution has linear communication, computation and GC storage complexity. As n becomes large these complexity parameters make this scheme undesirable, and more scalable key management schemes should be used.

In general, during the design of a multicast application, there are several issues that should be kept in consideration when choosing a key distribution scheme. We now provide an overview of some of these issues.

- **Dynamic nature of group membership:** It is important to efficiently handle members joining and leaving as this necessitates changes in the session key and possibly any intermediate keying information.

- **Ability to prevent member collusion:** No subset of the members should be able to collude and acquire future session keys or other member's key encrypting keys.

- **Scalability of the key distribution scheme:** In many applications the size of the group may be very large and possibly on the order of several million users. The required communication, storage, and computational resources should not become a hindrance to providing the service as the group size increases.

In Section 2.1, we summarized the work of [3, 4] for the distribution of secret information via broadcast messages. These results provide a insight into the communication resources needed to achieve the above goals. In particular, it was shown in Theorem 1 that for a key size of B bits, the message needed to update a group of n users must be at least nB bits to provide *perfect security* in the key distribution. One key result of [3] is that in order to achieve a smaller broadcast size, it is necessary to do away with the constraint that the private information held by each user is mutually independent. Therefore, to reduce the usage of communication resources, the users must share secret information.

One strategy for having users share secret information is to arrange the keys according to a tree structure. The tree based approach to group rekeying was originally presented by Wallner et al. [7], and independently by Wong et al. [8]. In such schemes an a-ary tree of depth $\log_a n$ is used to break the multicast group into hierarchical subgroups. Each member is assigned to a unique leaf of the tree. KEKs are associated with all of the tree nodes, including the root and leaf nodes. A member has knowledge of all KEKs from his leaf to the root node. Thus, some KEKs are shared by multiple users. Adding members to the group amounts to adding more depth to the tree [9], or adding new branches to the tree [8]. Upon member departure the session key and all the internal node KEKs assigned to that member become compromised and must be renewed. Due to the tree structure, the communication overhead is $\mathcal{O}(\log n)$, while the storage for the center is $\mathcal{O}(n)$ and for the receiver is $\mathcal{O}(\log n)$.

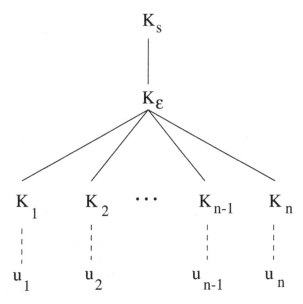

FIGURE 6.1. The basic key distribution scheme.

Various modifications to the tree scheme have been proposed. In [10], a modification to the scheme of Wallner et al. is presented. By using pseudorandom generators, their scheme reduces the usage of communication resources by a factor of two. Similarly, Balenson et al. [9] were able to reduce the communication requirements by a factor of two using one-way function trees. In [11] Canetti et al. examine the tradeoffs between storage and communication requirements, and a modification to the tree-based schemes of [7, 8] is presented that achieves sublinear server-side storage. Further, in [12], it was shown that the optimal key distribution for a group leads to Huffman trees and the average number of keys assigned to a member is related to the entropy of the statistics of the member deletion event.

6.1 A Basic Key Management Scheme

In this section, we present a simplified key management scheme that will be used in the discussions in Section 6.3.1 where we introduce an improved format for the rekeying message. The key management scheme presented here is an elementary key management scheme that consists of two layers of KEKs, and a SK that is used to protect the bulk content.

Consider a group of n multimedia users who will share a multimedia multicast. In the simple key distribution scheme for n users, depicted in Figure 6.1, user u_j has two key encrypting keys K_j and K_ϵ, and the session key K_s. The KEK K_ϵ is the root KEK and is used to encrypt messages that

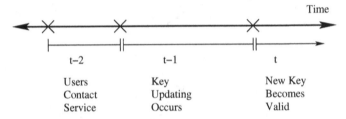

FIGURE 6.2. The time intervals $t-2$, $t-1$ and t. The joining/departing user contacts the service during time interval $t-2$, the rekeying messages are transmitted during $t-1$, and new key information takes effect at the beginning of time interval t.

update K_s. The remaining keys $K_1, K_2, \cdots, K_n$ are KEKs that are used to protect updates of K_ϵ.

Due to the dynamic nature of the group, and the possible expiration of keying material, it is necessary to update both the SK and KEKs using rekeying messages. The three operations involved are key refreshing, key updating when a new user joins the service, and key updating when a user departs the service. In the the discussions that follow, we use an integer-valued time index to denote the time intervals during which fundamental operations occur, and assume that there is a system-level mechanism that flags or synchronizes the users to the same time frame. We shall always use the time index t to denote the interval for which the new key information will become valid. Time interval $t-1$ will correspond to the time interval during which the new key information is being transmitted. Further, time interval $t-2$ corresponds to the interval of time during which a new member contacts the service provider wishing to join, or a current member announces to the service provider his desire to depart the service. We have depicted these cases in Figure 6.2. Observe that it is not necessary that the time intervals have the same duration.

6.1.1 Key Refreshing

Refreshing the session key is important in secure communication. As a session key is used, more information is released to an adversary, which increases the chance that a SK will be compromised. Therefore, periodic renewal of the session key is required in order to maintain a desired level of content protection. By renewing keying material in a secure manner, the effects of a session key compromise may be localized to a short period of data.

The cryptoperiod associated with a session key is governed by many application-specific considerations. First, the value of the data should be examined and the allowable amount of unprotected (compromised) data should be addressed. For example, the broadcast of a sporting event might allow the data to be unprotected for a short period, whereas a video conference between corporate executives would likely have stricter security requirements and necessitate more frequent key refreshing.

Since the amount of data encrypted using KEKs is usually much smaller than the amount of data encrypted by a session key, it is not necessary to refresh KEKs. Therefore, KEKs from the previous time interval $t-1$ carry over to the next time interval. In order to update the session key $K_s(t-1)$ to a new session key $K_s(t)$, the group center generates $K_s(t)$ and encrypts it using the root KEK $K_\epsilon(t)$. This produces a rekeying message $\alpha_s(t) = E_{K_\epsilon(t)}(K_s(t))$, where we use $E_K(m)$ to denote the encryption of m using the key K. The message $\alpha_s(t)$ is sent to the users.

6.1.2 Member Join

In multimedia services, such as pay-per-view and video conferences, the group membership will be dynamic. Members may want to join and depart the service. It is important to be able to add new members to any group in a manner that does not allow new members to have access to previous data. In a pay-per-view system, this amounts to ensuring that members can only watch what they pay for, while in a corporate video conference there might be sensitive material that is not appropriate for new members to know.

Suppose that, during time interval $t-2$, a new user contacts the service desiring to become a group member. If there were $n-1$ users at time $t-2$ then there will be n users at time t. During time interval $t-1$, the rekeying information must be distributed to the $n-1$ current members. Observe that we must renew both the SK and the root KEK in order to prevent the new user from accessing previous rekeying messages and to prevent access to prior content.

The first stage of the key updating procedure requires updating the root KEK from $K_\epsilon(t-1)$ to $K_\epsilon(t)$. Since all of the members at time $t-1$ share $K_\epsilon(t-1)$, the group center may communicate the new KEK $K_\epsilon(t)$ securely to these members by forming and transmitting the message $\alpha_\epsilon(t) = E_{K_\epsilon(t-1)}(K_\epsilon(t))$. Next, the service provider generates a new session key $K_s(t)$ and updates the session key using a rekeying message of the form $\alpha_s(t) = E_{K_\epsilon(t)}(K_s(t))$.

Meanwhile, during time interval $t-1$, the new user completes registration with the service and is given the new keys $K_s(t)$, $K_\epsilon(t)$, and K_{n+1}. This completes the actions required during time interval $t-1$, and at the start of time interval t all of the $n+1$ members have the new keying material.

6.1.3 Member Departure

Let us consider the case when user u_n leaves the group at time frame $t-1$. Since user u_n knows $K_s(t-1)$ and $K_\epsilon(t-1)$ these keys must be renewed. First K_ϵ is renewed. To accomplish this the GC forms a new key $K_\epsilon(t)$ and

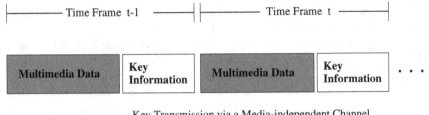

Key Transmission via a Media-independent Channel

Key Transmission via a Media-dependent Channel

FIGURE 6.3. Two approaches to distributing the key information in multimedia multicasting: (a) using a media-independent channel, and (b) using a media-dependent channel.

encrypts it with the keys K_j for $j \neq n$. A single message

$$\alpha_\epsilon(t) = E_{K_1}\left(K_\epsilon(t)\right) \| E_{K_2}\left(K_\epsilon(t)\right) \| \cdots \| E_{K_{n-1}}\left(K_\epsilon(t)\right) \qquad (6.1)$$

is formed and sent to all the users using either the media-independent or media-dependent channel. Here we use the symbol $\|$ to denote concatenation of bit streams. Next, the session key is updated. The GC forms a new SK $K_s(t)$ and encrypts using the new KEK $K_\epsilon(t)$ to form $\alpha_s(t) = E_{K_\epsilon(t)}\left(K_s(t)\right)$. This message is then sent to the users.

As a final note, we observe that the size of this message agrees with Theorem 1. Here, the message α consists of the concatenation of $n-1$ smaller messages. The message α is distributed to the $n-1$ remaining users. The total length of the message is $n-1$ times the block size of the encryption algorithm employed. Since the key length is smaller than the block size, we agree with Theorem 1.

6.2 Distribution of Rekeying Messages for Multimedia

After the formation of the rekeying messages, they must be delivered to the users. This issue is rarely considered in the secure multicast literature. However, it is an integral part of a system's design. For the transmission of multimedia data, we have identified two distinct classes of mechanisms, depicted in Figure 6.3, that are available for the delivery of the rekeying messages:

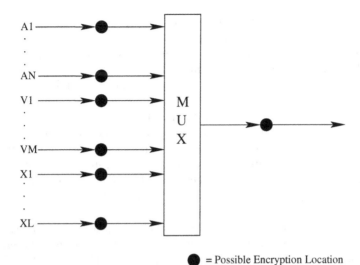

= Possible Encryption Location

FIGURE 6.4. A generic multiplexing diagram depicting several audio streams (A1 - AN), video streams (V1 - VM), and auxiliary streams (X1 - XL). Also depicted are locations where encryption is possible.

- **Media-Independent Channel:** In this mode, the rekeying messages are conveyed by a means totally disjoint from the multimedia content.

- **Media-Dependent Channel:** A media-dependent channel exists when the media is capable of having a small amount of data imperceptibly hidden inside the host media.

In a conventional, non-secure multimedia application, the multimedia data consists of multiple streams. Depending on the application, these streams may either be multiplexed together and placed onto the network, or treated as separate layers that are passed onto a separate delivery protocol. For example, in MPEG-2 Systems a multiplexer operation will multiplex the audio and video data into either a program stream or a transport stream [94]. As another example, MPEG-4 provides packetized elementary streams to the Delivery Multimedia Integration Framework (DMIF) which deals with different delivery scenarios and allows for desirable delivery techniques such as unequal error protection (UEP) [95,96].

The location of the encryption operation in a multimedia application's design, as well as the mode that the encryption operates under, has a significant effect on the performance of the multimedia multicast service. In Figure 6.4, we present a generic diagram that captures the multiplexing involved in H.324, MPEG-2 Systems program stream or transport stream, or the operations of the DMIF in MPEG-4. Several audio streams (A1 - AN), video streams (V1 - VM), and auxiliary streams (X1 - XL) are fed as input into the multiplexer. Upon output is a data stream that consists of packets that have been interlaced. With respect to this diagram, there are two locations where encryption can be placed. Encryption can either be located before the multiplexer or after it. If encryption is placed after the MUX, then there are two manners in which it can encrypt the data stream. First, it can encrypt each packet individually, thereby maintaining the separation of the packets that was introduced by the multiplexer. The second option is for the encryption to operate in a streaming mode, such as cipher block chaining [97], whereby the separation between different media packets will be lost. The disadvantage of the latter mode of operation is that it is no longer possible to treat the layers separately, which is essential to performing important delivery techniques like UEP. Therefore, if encryption is placed after the MUX, it should maintain the separation between the packets.

However, it is not necessary to place the encryption after the MUX to maintain the separation between the layers. In fact, placing the encryption before the MUX will encrypt each media or object stream independently, and the multiplexer will interleave the various encrypted streams into separated packets. The multiplexer and transmitter will then maintain the separation between the different media streams, which is essential for reliable delivery of multimedia. Therefore, there is no advantage in placing encryption after the MUX since the segregation between the different streams should be maintained, and hence encryption should be done prior to multiplexing. In fact, in the MPEG-4 IPMP framework IPMP control points are located prior to the DMIF at the encoder [98].

For the remainder of this section, we shall discuss the different mechanisms for distributing the rekeying messages. For each method we will discuss its advantages and disadvantages.

6.2.1 Media-Independent Channel

The first method to convey the rekeying information is to use a channel that is independent of the multimedia content. This can be accomplished in several different ways. First, one could have a security system that is completely separate from the multimedia system, and the key information is transmitted using any other channel that is available to the application. A second manner by which this can be accomplished is through a Systems level operation. In fact, MPEG-2 Systems (ISO/IEC 13818-1) provides the Entitlement Control Messages (ECM) as a means to convey keys associated with scrambling MPEG-2 multimedia streams. The ECM is transmitted as a stream separate from the multimedia. As another example, the MPEG-4 standard also provides a Systems level data stream to convey security information. In MPEG-4, the Intellectual Property Management and Protection (IPMP) framework provides IPMP Descriptors (IPMP-Ds) and IPMP Elementary Streams (IPMP-ESs) that can help an IPMP system decrypt or authenticate media elementary streams [98]. Both the MPEG-2 ECMs,

MPEG-4 IPMP-Ds and MPEG-4 IPMP-ESs can be used to convey rekeying information associated with a multicast service.

Further, many multimedia standards provide data fields that may be used by the system designers to convey non-normative application-specific parameters. For example, in MPEG-1 Video the bitstream format for the video sequence layer, the group of pictures layer, and the picture layer provides a mechanism to convey optional *user* data. These fields may be also used to convey security data, such as rekeying messages.

One of the advantages of using the media-independent channel is the ability to assign a delivery protocol to the rekeying messages that is different from the delivery protocols used by the other components of the data stream. Since encryption and decryption keys must be exactly known in order to perform decryption, rekeying messages are extremely sensitive to errors. It is essential that all receivers completely receive a correct rekeying message before the new key takes effect. Without a mechanism to ensure that a rekeying message is received by all legitimate members, some users will be unable to decrypt future content and future rekeying messages.

When the rekeying messages are transmitted using a media-independent channel, their delivery can be performed using a reliable multicasting protocol, such as RMTP and SRM [1, 99–101]. However, in addition to using reliable multicasting, it is necessary to add a feedback mechanism at the application layer. In a multicast security system, it is necessary that the server knows that all users have correctly received the rekeying message before proceeding to the next rekeying message or encrypting the service with the new session key. Therefore, before switching to the new key, the server must wait for an acknowledgement message from each of the clients announcing that they have successfully received the rekeying message.

The use of a media-independent channel can introduce a network security weakness even if there is no cryptographic weakness in the key management scheme. We illustrate this with the following example. When transmitting the rekeying messages in the media-independent mode, the keying messages will be in an encrypted format, such as depicted in (6.1), and kept separated from the other types of data packets. It is possible for an adversary to eavesdrop on the network and observe the presence of these rekeying messages. Even if the rekeying messages are further encrypted by the session key K_s, an eavesdropper on the network may simply observe the rekeying message substream to measure valuable statistical data regarding the multicast membership. For example, if an adversary knows that the key size used is 64 bits and that the rekeying message is of the form (6.1), then when he observes a rekeying message of 64000 bits, he may infer that there are 1000 users in the service. The leakage of statistical information regarding the service membership is a security flaw that can be addressed by using a media-dependent channel. In [102] other system weaknesses were identified that can occur in multicast key distribution schemes even when the underlying cryptographic algorithm is provably correct.

6.2.2   Media-Dependent Channel

A media-dependent channel exists when small amounts of information can be embedded invisibly in the data. In these cases, the rekeying information may be embedded in the content and distributed to those who receive the data [49, 103]. Data embedding, or digital steganography, techniques allow for an information signal to be *hidden* in another signal, known as the cover signal, without dramatically distorting the cover signal. Effective data embedding techniques are those that can invisibly embed data in the cover signal, allow for easy extraction of the embedded information, and achieve a high embedding rate.

Multimedia data types, such as speech, image, and video are well suited for embedding information since introducing a small amount of distortion in their waveforms does not significantly alter perceptual quality [104–106]. Generic data structures are not well suited for hiding information. The most popular purpose for data embedding is digital watermarking, in which ownership or copyright information is inserted in the cover signal. In this case, the embedding technique must also be robust to attempts to remove or destroy the watermark. Data embedding can also be used to convey side information, such as embedding messages in the content.

Many papers exist on embedding information and watermarks in video. In [105], Hartung and Girod describe a method for inserting digital watermarks into the compressed bitstream of MPEG-2 coded video. They found that they could embed a watermark of 1.25 to 125 bytes/second in NTSC signals. Another method for embedding information in video was presented in [107], and applied to distributing textual information in a video conferencing system. As another example of a scheme with a high embedding rate, a data embedding scheme that is compatible with standards such as H.263 and MPEG-2 was proposed in [108, 109]. This data embedding technique uses the fractional-pel motion vector as the cover signal for the embedded data, and is able to embed a high bitrate information signal into a video bitstream with an acceptable visual quality degradation. This method for data embedding will be used later in this chapter to demonstrate the feasibility of our multimedia multicast key distribution philosophy.

Associated with many embedding schemes is an *embedding key* which governs how the information is embedded into the cover signal. The size of the embedding key dictates the difficulty for an adversary to attack the embedding rule. For example, in [108, 109], 2 bits of information can be embedded per macroblock, and these 2 bits are embedded by mapping the motion vector to one of 4 regions. There are $4! = 24$ different embedding rules possible. We may therefore associate a 5-bit embedding key K_{emb} with one of these 24 different methods. If a user has the key associated with how the data was embedded, then he may extract the information signal in the multimedia data. An adversary, however, would have to search these 24 possibilities to determine the correct embedding rule to extract the embedded information.

It is desirable to have the size of K_{emb} large in order to make it difficult for an adversary to attack the embedding rule. We now describe the method by which we extend the embedding key size of [108, 109] for use in our later simulations. Suppose that we break the information we wish to embed into 2-bit chunks c_j. We shall choose security parameter q that is a non-negative integer. At random, we shall choose q different embedding rules $(r_0, r_1, \cdots, r_{q-1})$, allowing for repetition in the rules selected. Each embedding rule r_k describes one of the 24 possible ways to map 2 bits to 4 regions. We assign an embedding rule r_k for each chunk c_j according to $k \equiv j \pmod{q}$. Thus, the $0, q, 2q, \cdots$ 2-bit chunks use embedding rule r_0, the $1, q+1, 2q+1, \cdots$ 2-bit chunks use embedding rule r_1, and so on. The embedding key is thus the concatenation of these rules, which is a key space of 24^q possibilities, and requires $\lceil q \log_2 24 \rceil$ bits to represent. For example, choosing $q = 12$ yields an embedding key size of 56 bits.

The rekeying messages used in either the media-independent or media-dependent cases are almost identical. When using the media-independent approach, only the information needed to update the SK and KEKs needs to be transmitted. However, when using a media-dependent approach, the embedding key must also be updated, requiring that an additional rekeying operation is performed.

The primary advantage of using data embedding to convey rekeying messages compared to a media-independent channel is that data embedding provides an additional layer of security that hides the presence of rekeying messages from potential adversaries. In the conventional approach of using a media-independent channel to convey the rekeying messages, an adversary can observe the external channel and determine information about the membership dynamics of the multicast service, such as the rate at which members join and leave the service as well as being able to infer information about the group membership. From a security point of view, this provides valuable information to a potential adversary. In comparison, data embedding provides *covert* information transferral, whereby the bit rate of the multimedia source is maintained and it is impossible for an eavesdropper to measure information regarding the occurrence of a rekeying operation.

Another effect of the additional layer of security provided by data embedding is the introduction of the embedding key, which must also be maintained by the service provider and stored by the user. A positive benefit of this is that an adversary will not only have to attack the SK and KEKs, but he will also have to attack the key governing the embedding rule in order to acquire rekeying messages. Since the rekeying message is embedded into the multimedia, it is encrypted by the SK, and thereby protected by the SK, the KEK, and the embedding key. For this reason, it is therefore important that the key length of the embedding key is sufficiently long to make it difficult for the adversary to search the embedding key space. We note that a similar increase in protection can be achieved in the media-independent channel by increasing the key length of the session key or by introducing an additional SK. However, encryption algorithms are typically designed for a small set of specified key lengths [13] and it might not be possible to increase the length of the session key.

Finally, when using a media-dependent channel, it is possible to maintain the original data rate of the media without performing the additional computations associated with transcoding. When using a media-independent channel, the rekeying messages introduce additional communication overhead that is in addition to the bandwidth needed to convey the media. In order to keep the data rate of media and rekeying messages identical to the data rate of just the original media, it is necessary to perform computationally intensive transcoding of the media to a lower data rate. However, when using a media-dependent approach to conveying the rekeying messages, the original data rate is maintained, and the data embedding operation provides a graceful degradation of media quality as more data is embedded.

When using media-dependent channels, the issue of reliability becomes more pronounced than in the media-indpendent case since it is not possible to send the rekeying messages through a delivery mechanism separate from the multimedia data. Since multimedia data is delay sensitive and often transmitted on error-prone channels using *best effort* delivery protocols, it is likely that some media packets will be lost, and the rendering buffer will be filled using the data that successfully arrives. However, when using a media-dependent channel, the lost media packets might contain part of a rekeying message. Since the rekeying messages are embedded in multimedia, which is being delivered through best effort delivery protocols, it is not possible to apply delivery protocols employing retransmissions to improve the reliability of key delivery. There is therefore a tradeoff between covert information transferral and reliability in delivering the rekeying message.

We noted earlier that it is important that the rekeying message is completely received by all users before using the new key. We may, however, address the reliable delivery of the rekeying messages at the application layer. For example, the multimedia system may employ a centralized error recovery technique similar to the NP protocol of [110], however operating at the application layer. The server application takes the k data packets corresponding to a rekeying message and would form h additional parity packets. These $k + h$ packets would be used transmitted as the rekeying message that would be transferred through the media-dependent channel. At the completion of sending the $k + h$ packets, the server would send a message polling the clients whether they were able to successfully decode the rekeying message. The clients would send back acknowledgement messages to the server. If not all of the clients were able to receive the complete rekeying message, the server would employ retransmission, and the process would repeat until all users have successfully received the rekeying message.

When all users have received the rekeying message, the server would issue a message instructing all users that it is appropriate to use the new key.

6.3   An Improved Rekeying Message Format

We have described how a rekeying message can be formed during member departure so that each of the remaining members can receive the new root KEK (key encrypting key) by decrypting an appropriate segment of the message using their private KEK. In practice this requires sending additional information that flags to all of the users which segment belongs to which user. Not only does this mean that additional communication overhead is required, but also that sensitive information regarding user identities is released. In particular, adversaries who are members of the service can collect information about other keying messages intended for other users. In order to circumvent this potential weakness, we propose a new format for the rekeying message that is a single, homogenized message from which each user may extract the new root KEK. Such an approach has the advantage that user-specific keying information is not available to other users.

The problem of distributing information simultaneously to multiple users via a single broadcast message while maintaining user anonymity has been previously studied in the literature. Just et al. [3] and Blundo et al. [4] each present a method using polynomial interpolation whereby the broadcast message does not have a partitioned structure like the message in (6.1). A drawback of both of these schemes is that they are suitable for only one transmission, and are not reusable. Specifically, when used to distribute identical information to multiple recipients, each user's secret information is valid for only one transmission, and then is available for other group members to acquire. This is a problem since members may acquire other user's secret information and use this knowledge to enjoy the service after they cancel their membership. In order to use these schemes when the keying material must be updated multiple times, it is necessary to distribute to each user enough copies of private material to cover the amount of updates needed. Thus, although these schemes use a composite message structure and don't require additional communication overhead for flagging the users, they are not appropriate for applications that require recurrent key distribution.

We therefore desire a scheme that allows for private keying material to be reused while providing a homogenized message form. In Section 6.3.1, we shall describe a new message format that makes use of one-way functions and a broadcast seed to protect each user's private information from compromise [49]. Additionally, although our use of one-way functions can be applied to the polynomial interpolation methods of [3, 4], our message format only requires the use of the basic operations of large integer multiplication and modular arithmetic, and does not require the additional functions needed to calculate interpolating polynomials. Then, in Section 6.3.3, we describe how our message format would be used in a tree-based key management scheme to achieve logarithmic usage of communication resources.

First, we introduce parametric (or keyed) one-way functions [97, 111], which are the building blocks of our message form.

Definition 1. *A parametric one-way function (POWF) h is a function from $\mathcal{X} \times \mathcal{Y} \to \mathcal{Z}$ such that given $z = h(x, y)$ and y it is computationally difficult to determine x.*

Parametric one-way functions are families of one-way functions [13, 97] that are parameterized by the parameter y. The discrete logarithm provides an example of a POWF since if p is a large prime, and x and y non-identity elements of Z_p^*, the multiplicative subgroup of integers modulo p, it is computationally difficult to determine x given $z = y^x \pmod{p}$ and y [13, 97]. Since symmetric ciphers are typically computationally efficient compared to one-way functions that employ modular exponentiation, practical one-way functions should be implemented by means of a symmetric encryption cipher. For example, if we let g be a suitable hash function, and E_x a symmetric cipher, then $h(x, y) = g(E_x(y))$ is a POWF. In this case, only ciphers that are secure against known plaintext attacks [97], such as DES or Rijndael, are appropriate. Further, we note that it is not necessary that the hash function g have any cryptographic properties since the required strength is provided by E. Throughout this chapter we shall assume the existence of POWFs that map sequences of $2B$ bits into sequences of B bits.

6.3.1   Basic Message Form

For the basic message form, we shall use the key distribution scheme depicted in Figure 6.1. Suppose that at time $t-2$ the group consists of n users $u_1, u_2, \cdots, u_n$. Each user u_i has a personal B-bit KEK K_i that is known only by the group center and user u_i. Additionally, all of the users share a B-bit root KEK and a session key that will vary with time.

The group center makes available a POWF h that maps a sequence (x, y) of $2B$ bits to B bits. A new function f is defined by prepending a single 1 bit in front of the output of $h(x, y)$, that is $f(x, y) = 1 \| h(x, y)$. The purpose of prepending a bit is to ensure that the modulo operation used by each user will yield $K_\epsilon(t)$.

Suppose, without loss of generality, that user n decides to leave at time $t - 2$, then both $K_\epsilon(t - 1)$ and $K_s(t - 1)$ must be updated. The root KEK is updated first, and then used to encrypt the new session key. In order to update $K_\epsilon(t - 1)$, the GC first broadcasts a B-bit random seed $\mu(t)$. Next, the GC forms $K_\epsilon(t)$ and calculates the rekeying message as

$$\alpha_\epsilon(t) = K_\epsilon(t) + \prod_{i=1}^{n-1} f\big(K_i, \mu(t)\big).$$ (6.2)

A legitimate member u_i may decode $\alpha_\epsilon(t)$ to get the key $K_\epsilon(t)$ by calculating $\alpha_\epsilon(t)$ (mod $f(K_i, \mu(t))$).

We observe that the only property of $\mu(t)$ that is needed is that it is known by all of the recipients. We can therefore achieve a different variation of the scheme by choosing $\mu(t) = K_\epsilon(t-1)$ or $\mu(t) = K_s(t-1)$, which does not require the transmission of the random seed by the system.

We now discuss how this message format reduces the communication overhead compared to a partitioned message format, such as is depicted in (6.1). Current multicast key management schemes, such as [7, 8, 10], focus on the size of the payload (the rekeying information), and not on the size of the entire message (including the rekeying message and the header). In fact, the transmission of the messages that flag the users which portion of the message is intended for them can add significant communication overhead when used in conventional tree-based schemes. To illustrate this, we consider the basic key management scheme depicted in Figure 6.1, with $n+1$ users. When using the partitioned message form of (6.1), it is necessary to send a header message that describes the user IDs associated with each of the blocks in the payload rekeying message. Since it requires at least $\log_2(n)$ bits to describe the user IDs for n users, we need an additional overhead of $n\log_2(n)$. Therefore, the percentage of the message size that corresponds to the communication overhead is

$$\rho = \frac{n \log_2 n}{nB_k + n \log_2 n},$$ (6.3)

where B_K is the bit length of the KEK K_ϵ. For large n the communication becomes a significant portion of the message size.

However, the message format of (6.2) is a single, homogenized message that does not require any communication overhead. If we use $\mu(t) = K_\epsilon(t-1)$, then it is not necessary to broadcast $\mu(t)$ and the total message size of (6.2) is $n(B_K + 1)$, whereas the total message size from the traditional format was $n(B_K + \log_2 n)$. Therefore, as long as $\log_2 n > 1$, the message format of (6.2) is more efficient in terms of communication. This occurs when we are providing service to a group with more than 2 users. Therefore, we have made a tradeoff between communication and computation. The message format of (6.2) uses less overall communication at an expense of requiring more computation to form the message.

6.3.2 Security Analysis of Residue-based Method

The residue-based method for multicast key distribution was described in Section 6.3. The basic form of rekeying message in the residue-based method is

$$\alpha = X + \prod_{j=1}^{r} Y_j, \tag{6.4}$$

where X and Y_j are drawn uniformly from the set $\{1, 2, \cdots, B\}$. The variable X corresponds to the secret, or the key, that is being convey, while the Y_j are the user-specific shares that mask the secret.

This section describes an information theoretic investigation of the security that this method provides for protecting X, and some motivation for using one-way functions with a time-varying seed.

Information Theoretic Analysis

Consider the scenario where there is only one Y term, and define the random variable $Z = X + Y$. In general, the entropy of Z is difficult to relate to the entropy of two arbitrary random variables X and Y. The following relationship holds

$$H(Z) \le H(X, Y) \le H(X) + H(Y), \tag{6.5}$$

and in general the bound need not be met. In particular, the consider the random variables

$$X = -Y = \begin{cases} 1 & : \quad \text{with probability } p(1) = 1/2 \\ 0 & : \quad \text{with probability } p(0) = 1/2 \end{cases} \tag{6.6}$$

In this example, $H(X) = H(Y) = 1$, while $Z = 0$ so that $H(Z) = 0$. The following lemma places a lower bound on the entropy of Z.

Lemma 8. *Suppose that* $Z = X + Y$, *then* $H(Z|X) = H(Y|X)$.

Proof.

$$
\begin{aligned}
H(Z|X) &= \sum_{x} p(x) H(Z|X = x) \\
&= -\sum_{x} p(x) \sum_{z} p(Z = z|X = x) \log p(Z = z|X = x) \\
&= -\sum_{x} p(x) \sum_{y} p(Y = z - x|X = x) \log p(Y = z - x|X = x) \\
&= \sum_{x} p(x) H(Y|X = x) \\
&= H(Y|X).
\end{aligned}
$$

$\square$

Thus, $H(Z) \ge H(Z|X) = H(Y|X) = H(Y)$, since X and Y are assumed independent. Similarly, $H(Z) \ge H(X)$.

The pdf of $Z = X + Y$ is simply the convolution of the pdf of X and the pdf of Y. Now, when X and Y are uniformly drawn from $\{1, \cdots, B\}$, then the pdf of Z is a triangular function, and the entropy of Z may be calculated directly. Suppose that $h(k) = B^2 p_Z(k)$, then the entropy of Z is

$$
H(Z) = -\frac{1}{B^2}\left[\left(\sum_{k=2}^{2B} h(k)\log h(k)\right) - \left(2\log B \sum h(k)\right)\right] \quad (6.7)
$$

$$
= -\frac{1}{B^2}\left[\sum h(k)\log h(k)\right] + \frac{2\log B}{B^2}\sum h(k) \quad (6.8)
$$

$$
= -\frac{1}{B^2}\left[\sum h(k)\log h(k)\right] + 2\log B. \quad (6.9)
$$

Due to the symmetry of the triangle function $h(k)$,

$$
\sum_{k=2}^{2B} h(k)\log h(k) = 2\left(\sum_{j=1}^{B-1} j\log j\right) + B\log B. \quad (6.10)
$$

Thus

$$
H(Z) = -\frac{2}{B^2}\left(\sum_{j=1}^{B-1} j\log j\right) - \frac{1}{B}\log B + 2\log B \quad (6.11)
$$

$$
= \left(2 - \frac{1}{B}\right)\log B - \frac{2}{B^2}\left(\sum_{j=1}^{B-1} j\log j\right). \quad (6.12)
$$

The entropy of the sum was calculated for X and Y drawn uniformly from a range $\{1, 2, \cdots, B\}$, where $B = 2^b$ and is recorded in Table 6.1. Examining this table reveals that the difference between the entropy in Z and the entropy of either X or Y tends toward an asymptotic limit. The exact value and significance of this limit is currently not known.

The security of the residue-based method is measured by the uncertainty that remains in the key given only the observation of the rekeying message. For the case of a single term in the product, this is measured by the entropy $H(X|Z)$. The following lemma relates $H(X|Z)$ to the entropies $H(X)$, $H(Y)$, and $H(Z)$.

Lemma 9. *Suppose X and Y are independent, and $Z = X + Y$, then*

$$
H(Z|X) = H(X) + H(Y) - H(Z). \quad (6.13)
$$

Proof. By application of the chain rule, $H(X|Z) = H(X,Z) - H(Z)$. Observe that there is a unique correspondence between the joint variable (X, Z) and (X, Y). Therefore, $H(X, Z) = H(X, Y) = H(X) + H(Y)$. Substitution gives the desired result.   □

TABLE 6.1. The entropy of the sum $Z = X + Y$, where X and Y are drawn uniformly from integers between 1 and $B = 2^b$.

b	H(Z)	b	H(Z)
1	1.5000000000000	13	13.7213474774632
2	2.6556390622296	14	14.7213475090783
3	3.7023191426459	15	15.7213475174478
4	4.7159395672686	16	16.7213475196565
5	5.7198327831914	17	17.7213475202379
6	6.7209281467435	18	18.7213475203902
7	7.7212325045380	19	19.7213475204300
8	8.7213162233391	20	20.7213475204415
9	9.7213390603855	21	21.7213475204434
10	10.7213452464840	22	22.7213475204440
11	11.7213469122180	23	23.7213475204448
12	12.7213473584537	24	24.7213475204435

Applying this lemma with the values presented in Table 6.1 gives that $H(X|Z) \approx H(X) - 0.721347$. This result implies that roughly one bit of security is lost when there is only a single Y term. Thus, an adversary must only search a keyspace that is half as large as the original keyspace.

If the original keyspace is sufficiently large, then this reduction might not be significant. For example, searching a keyspace of 100 bits is effectively as difficult as searching a keyspace of 99 bits.

The security of this scheme when more Y_j terms are used remains to be investigated. It is conjectured that the amount of bits lost will increase since the distribution of $Y = Y_1 Y_2 \cdots Y_r$ will no longer be uniform. Additionally, the exact value of the 0.721347 term remains to be explored. Finally, since direct calculation of the entropies for large $B \approx 2^{20}$ takes considerable computing effort, it would be desirable to construct bounds on the entropy of $Y = Y_1 Y_2 \cdots Y_r$, as well as on the entropy of $Z = X + Y$.

Attacks by Insiders

We now examine the possibility for a member of the group to attack the security of the system by gathering or inferring information not intended for them. Suppose that the basic form of the rekeying message is

$$\alpha(t) = X(t) + Y = X(t) + \prod_{j=1}^{r} Y_j, \tag{6.14}$$

where $X(t)$ denotes the time-varying key that the GC is distributing to the users. The Y_j are user specific secrets that allow a user to determine $X(t)$ given α by performing a modulo operation. In a dynamic environment, it is important to prevent members from acquiring other member's secrets.

In the basic form of the rekeying message, once the user u_s has determined X, he may determine $\prod_{j \neq s} Y_j$ by

$$\prod_{j \neq s} Y_j = \frac{\alpha(t) - X(t)}{Y_s}. \tag{6.15}$$

This allows user u_s to depart the system, receive a future rekeying $\alpha(t)$, and use $\prod_{j \neq s} Y_j$ in the modulo operation to determine future $X(t)$.

It is therefore necessary to make the Y term also time-varying in order to make it more difficult to acquire $X(t)$. An initial approach to solve this problem was to define $Y(t) = \lambda(t) \prod_{j=1}^{r} Y_j$. In this case, an inside adversary is able to calculate

$$A(t) = \lambda(t) \prod_{j=1}^{r} Y_j \tag{6.16}$$

Since the $\lambda(t)$ are chosen at random, one might expect that this would introduce enough randomness to make calculating $\prod_{j=1}^{r} Y_j$ difficult. This, however, is not the case.

Consider the probability that two random integers are relatively prime. A non-rigorous derivation of this probability would proceed as follows. The probability that one number is divisible by a prime p_i is $1/p_i$. If both numbers were chosen independently of each other, the probability that both are divisible by p_i is $1/p_i^2$ and hence the probability that they are not both divisible by p_i is $(1 - 1/p_i^2)$. The probability that two numbers are coprime can then be estimated by

$$W_2 = \prod_{p_i} \left(1 - \frac{1}{p_i^2}\right). \tag{6.17}$$

In order to calculate W_2 it is easier to start with $1/W_2$.

$$\frac{1}{W_2} = \prod_{p_i} \left(\frac{1}{1 - \frac{1}{p_i^2}}\right). \tag{6.18}$$

By expanding $\frac{1}{1-x}$ into a series expansion and observing that the product results in a term for every integer, we get

$$\frac{1}{W_2} = \sum_{n=1}^{\infty} \frac{1}{n^2} = \zeta(2) = \frac{\pi^2}{6} \tag{6.19}$$

where ζ is Riemann's zeta function [112]. Therefore, the probability of two numbers being relatively prime is $W_2 = \frac{6}{\pi^2}$. This can be extended to the probability that s numbers are relatively prime. Let W_s be the probability that s non-negative integers are relatively prime, then by the same idea as before

$$W_s = \prod_{p_i} \left(1 - \frac{1}{p_i^s}\right) \tag{6.20}$$

TABLE 6.2. Probabilities of Coprimality

W_2	0.608
W_3	0.832
W_4	0.924
W_5	0.964
W_6	0.983

which leads to

$$\frac{1}{W_s} = \sum_{n=1}^{\infty} \frac{1}{n^s} = \zeta(s). \tag{6.21}$$

We tabulate the first few such probabilities in Table 6.2.

This result states that an inside adversary, after gathering m observations $A(t_m)$ has an increasingly likely chance of calculating $\prod_{j=1}^{r} Y_j$. In fact, with just 8 observations, there is over a 99.5% chance that he will be able to calculate $\prod_{j=1}^{r} Y_j$.

In a dynamic scenario, calculating $\prod_{j \neq s} Y_j$ does not guarantee being able to acquire future $X(t)$. However, the Y_j of any other member who remains in the service will do. Unlike other cryptographic methods, such as RSA, where the factors have greater than 500 bits, each Y_j is typically less than a couple hundred bits and factoring the product $\prod_{j=1}^{r} Y_j$ will not be too difficult [13, 97]. In this case, the adversary's task is to reconstruct a Y_j given a list of factors of $\prod_{j=1}^{r} Y_j$. Since there is no guarantee how many factors each Y_j will have, it is not reasonable to rely on the difficulty of recombining factors to protect the key $X(t)$.

Other approaches to making the Y term time-varying have been examined. In order for the time-varying form $Y(t)$ to be secure, it must be difficult to calculate an individual Y_j term given knowledge of the rekeying message $\alpha(t)$ and the key $X(t)$. A natural approach for making it difficult to calculate a value y given a value $g(y)$ is to make g a one-way function. The idea of using one-way functions provides the difficulty needed to prevent an inside member from calculating another user's private key Y_j. However, using one-way functions alone did not provide security for protecting $X(t)$. As mentioned earlier, it is necessary to make the Y term time-varying, and based on the arguments presented above, one needs to make the product term $\prod_{j=1}^{r} Y_j$ time-varying. This was accomplished by introducing the time-varying broadcast seed.

By broadcasting $\mu(t)$ and using a non-reversible function f, the adversary is instead able to calculate

$$A_i = \prod_{j \neq i} f(K_j, \mu(t)). \tag{6.22}$$

Factoring A_i provides information about $f(K_j, \mu(t))$. Since it is difficult to acquire K_j given $\mu(t)$ and $f(K_j, \mu(t))$, the private user information is protected. At the next time instant, when $\mu(t+1)$ is broadcast, the adversary's

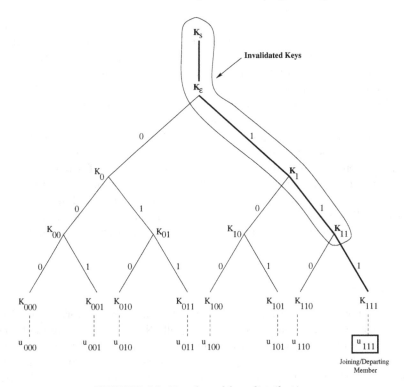

FIGURE 6.5. Tree-based key distribution.

knowledge of $f(K_j, \mu(t))$ does not help him in calculating $f(K_j, \mu(t+1))$, and he can extract K_ϵ only if he has the needed keys assigned to him.

6.3.3 Achieving Scalability

When the multicasting group is very large, it is necessary to make efficient usage of communication resources. Improved resource scalability can be achieved by employing a tree-based key management scheme to update the SK and KEKs [7, 8].

A binary tree is shown in Figure 6.5, though in the general case the tree can be an a-degree tree. Attached to the tree above the root node is the session key K_s. Each node in the tree is assigned a KEK called an internal key (IK) which is indexed by the path leading to itself. The symbol ϵ is used to denote empty string, which is the path of the root node to itself. Each user is assigned to a leaf and is given the IKs of the nodes from the leaf to the root node in addition to the session key. For example, user u_{111} is assigned keys K_{111}, K_{11}, K_1, K_ϵ, and K_s. All of the keys, with the possible exception of the leaf keys, may vary with time to reflect the changing dynamics of the group membership.

During periodic refreshes, only the session key needs to be updated, and the same protocol as presented in Section 6.1.1 can be used. We will now address how to operate during additions and deletions of members.

The GC is in charge of keeping track of the group members, and assigning them to positions on the tree. Although it is easiest to have the membership tree be a balanced tree, it is not necessary. For example, in [9], a non-balanced tree employing one-way functions is used in a key management scheme allowing member joins and departures is used. In this work, we shall just describe the procedure for adding members to a non-full balanced tree, and removing members from a full balanced tree. If a balanced tree is full, meaning all of the leaf nodes have members associated with them, then it is necessary to spawn a new layer of nodes when adding members. Additionally, by following the example of Balenson et al. [9] one can see how to make an approach handle member joins and departures for non-balanced trees.

Member Join

The member join operation does not involve the message format of (6.2) since each node of the key tree updates itself. Nonetheless, we present this case for completeness. Consider the binary tree depicted in Figure 6.5, that has 7 members u_{000} through u_{110}. If user u_{111} would like to join the group, the keys on the path from his leaf node to the tree's root as well as the SK, must be changed in order to prevent access to previous communications. Thus new $K_\epsilon(t)$, $K_1(t)$, $K_{11}(t)$ and $K_s(t)$ must be generated by the GC. The key encrypting keys can be updated from top to bottom by using $K_\epsilon(t-1)$ to encrypt $K_\epsilon(t)$, $K_1(t-1)$ to encrypt $K_1(t)$, and $K_{11}(t-1)$ to encrypt $K_{11}(t)$. Thus, all users can acquire the new root KEK, while only members u_{100}, u_{101}, and u_{110} can acquire $K_1(t)$. After updating the KEKs, the session key is updated by encrypting with the new root KEK $K_\epsilon(t)$.

Member Departure

When a member leaves the group, multiple keys become invalidated because that user shares these keys with other users. For example, in Figure 6.5, user u_{111} shares K_{11} with user u_{110}. Thus, if user u_{111} departs the multicast group, the key encrypting keys K_{11}, K_1, and K_ϵ become invalidated. These keys must be updated. Observe that K_{111} does not need to be updated since it is a private key and is not shared with any other users.

There are two basic approaches to updating the keys during a member departure: update the keys from the root node to leaf nodes, or from leaf nodes to root node. In the first approach, the *top-down* approach, when user u_{111} departs, the keys are updated in the order K_ϵ, K_1, and K_{11}. The second approach, the *bottom-up* approach, updates the keys in the order K_{11}, K_1, and K_ϵ. After updating the key encrypting keys, the root KEK

$K_\epsilon(t)$ can be used to encrypt the new session key $K_s(t)$ and a single message may be broadcast to all members.

Let us focus on how to update these keys using the top-down approach in conjunction with the new message form when user u_{111} departs. First, a random seed $\mu(t)$ is broadcast to all members, or some shared information, such as $K_\epsilon(t-1)$ is used as $\mu(t)$. Next, the root KEK $K_\epsilon(t-1)$ will be updated. In order to do this, the message

$$\alpha_\epsilon(t) = K_\epsilon(t) + f(K_0(t-1), \mu(t))f(K_{10}(t-1), \mu(t))f(K_{110}, \mu(t)) \quad (6.23)$$

is formed and broadcast. Next, $K_1(t-1)$ is updated by forming the message

$$\alpha_1(t) = K_1(t) + f(K_{10}(t-1), \mu(t))f(K_{110}(t-1), \mu(t)) \quad (6.24)$$

and broadcasting. The last KEK to update is $K_{11}(t-1)$. This can be done by sending the message

$$\alpha_{11}(t) = K_{11}(t) + f(K_{110}(t-1), \mu(t)). \quad (6.25)$$

Upon updating the KEKs, the session key may then be updated. To do this, the root KEK is used to encrypt $K_s(t)$ and the resulting message is broadcast.

In order to update the keys from a bottom-up approach, the random seed is broadcast, and then $K_{11}(t-1)$ is updated via

$$\alpha_{11}(t) = K_{11}(t) + \prod_{j=0}^{0} f(K_{11j}(t-1), \mu(t)) \quad (6.26)$$

The next key that is updated is $K_1(t-1)$. Since the two users beneath K_1 share a common key that is not invalidated by the departure of member u_{111}, we may reduce communication and computation by using this key to update K_1. The resulting message

$$\alpha_1(t) = K_1(t) + \prod_{j=0}^{1} f(K_{1j}(t), \mu(t)) \quad (6.27)$$

is broadcast. Since $K_{10}(t-1)$ is still valid, we implicitly updated $K_{10}(t) = K_{10}(t-1)$. To update $K_\epsilon(t-1)$ we may use the new key $K_1(t)$ as well as the old key $K_0(t) = K_0(t-1)$ and form the message

$$\alpha_\epsilon(t) = K_\epsilon(t) + \prod_{j=0}^{1} f(K_j(t), \mu(t)). \quad (6.28)$$

Finally, the session key is updated by encrypting the new session key $K_s(t)$ using the new root KEK $K_\epsilon(t)$, and broadcasting the message

$$\alpha_s(t) = E_{K_\epsilon(t)}(K_s(t)). \quad (6.29)$$

The amount of multiplications as well as the communication requirements needed to update all of the KEKs using the top-down approach and the bottom-up approach will differ. Assume that we have n users and keys assigned to each of these users using an a-ary tree. If the tree is a full, balanced tree with $L = \log_a n$ levels, then the amount of multiplications needed to update the KEKs during a member departure using a top-down approach is:

$$
\begin{aligned}
C_{td} &= \sum_{i=1}^{L} i(a-1) \\
&= (a-1)\frac{\log_a n(\log_a n + 1)}{2}.
\end{aligned}
\tag{6.30}
$$

Similarly, the amount of multiplications needed to update the KEKs using a bottom-up approach is

$$
\begin{aligned}
C_{bu} &= aL - 1 \\
&= a\log_a n - 1.
\end{aligned}
\tag{6.31}
$$

The amount of communication needed for each of these schemes is directly related to the amount of multiplications performed. If each internal key is B bits long, and a rekeying message requires M multiplications, then the message size will be $M(B+1)$ bits. Therefore, the bottom-up approach to renewing the keys requires less computation and communication. However, if the SK needs to be updated sooner, one may wish to use a top-down approach since it allows one to update the root KEK first, the session key next, and finally the remaining IKs.

6.4   System Feasibility Study

In this section, we study the issues related to the feasibility of using a key management system for multicast multimedia. When designing a cost effective system, one must consider the balance between computation, communication, and storage resources.

One of the primary advantages for using a tree-based key distribution scheme is that it achieves good scalability in the amount of communication needed to update the network. The need for using a tree-based key distribution scheme becomes more pronounced as the group size increases. If the group size is small, for example less than 10 users, there might not be any benefit from using a tree-based key distribution scheme, and one might want to consider the simple key distribution scheme presented in Section 6.3. However, the $\mathcal{O}(\log n)$ communication needed by most tree-based schemes makes the use of a tree-based scheme essential when the group size is several thousand or more users.

Another issue that should be considered is the amount of storage needed by the GC and each individual user. If each user has extremely limited storage, then the simple distribution scheme of Section 6.3 might be appropriate. However, although a tree-based scheme may require more storage for each user, and a factor more storage for the GC, typically this is not as important of a consideration as communication resources.

As an example, in the scheme presented in Section 6.3.3, the amount of multiplications (computation) needed to update the KEKs for the bottom-up approach was calculated to be $C_{bu} = a \log_a n - 1$. The communication needed is proportional to the amount of computation needed. The amount of storage needed by the GC to keep track of the KEKs is

$$S = \frac{a^{L+1} - 1}{a - 1} \tag{6.32}$$

keys, while the amount of storage needed by each user is $\log_a n + 2$ keys.

Next, one must consider the channel that one is transmitting the keys across. Whether transmitting via an external channel or an internal channel, there is a channel rate that governs how quickly the keying information may be distributed. For example, suppose we are transmitting the rekeying information for the scheme of Section 6.3.3 via an internal channel. If we denote R as the embeddable channel rate (in bits/second), B_{KEK} to be the key length of a KEK, B_s to be the key length of the session key, B_μ the bit length of the random seed $\mu(t)$, and B_{emb} to be the key length governing the data embedding rule, then the amount of time needed to update the entire system of keys is

$$T = \frac{C_{bu} B_{KEK} + B_s + B_{emb} + B_\mu}{R}. \tag{6.33}$$

Since T is related to the bit size of each of the keys, it is therefore related to the security levels protecting the service. This amount of time corresponds to the amount of time the departing member may still enjoy the service before no longer being able to decode the video stream. If we desire to increase the level of protection of the multimedia, then B_s must be increased, which leads to an increase in the amount of time needed to refresh the entire set of keys. Similarly, if we desire to increase the difficulty an adversary would have in decoding rekeying messages, then we need to increase B_{KEK}, which would also increase T.

In designing a system, these tradeoffs must be weighed and considered from a realistic point of view. Although it might be desirable to have extreme protection of the content, in a dynamic group, it is not realistic that it take an hour to update the set of keys.

To demonstrate these considerations, we present some simulation results using the data embedding scheme proposed in [109]. The degradation of the visual quality when different amounts of bits embedded per frame were measured for the *Foreman* and *Miss America* QCIF video sequences. The

H.263 TMN-11 video codec was used with annexes D, I, J, F turned on [113]. The bitrate in the simulation is 64kbps with a frame rate 10f/s, and every 12th frame is INTRA coded. The peak signal-to-noise ratio (PSNR) of luminance component with different data embedding rates are compared with the PSNR of luminance without embedding. In the simulations, the four cases compared correspond to when the number of bits embedded in a P-frame is upper bounded by 20, 40, 60 and no constraint (maximal). The PSNR differences are shown in Figure 6.6(a) for *Foreman* and Figure 6.6(b) for *Miss America*. Their average PSNR differences are also listed in Table 6.3. In all cases, the PSNR degradation of Luminance is within 1dB for both *Foreman* and *Miss America*, which normally cannot be detected by human visual system for video applications. Additionally, it was shown in [109] that data embedding at half-pel motion estimation at most degenerates the video coding performance back to integer-pel motion estimation without data embedding.

Using this data embedding scheme in conjunction with the bottom-up approach to member departure discussed earlier, we calculated the amount of time needed to refresh the entire network of keys for a tree of degree $a = 2$, and $n = 2^{20}$ or roughly one million users. We took $B_{KEK} = 56$ bits, $B_s = 56$ bits, $B_\mu = 56$ and $B_{emb} = 20$ bits as the bit lengths for the various keys. These values for B_{KEK}, B_s and B_μ were chosen since they correspond to the key size of the popular block cipher *DES*. The resulting times needed to refresh the keys are presented in Figure 6.7. The curves illustrate the inverse relationship with the amount of bits embedded per frame. Using these curves, one can determine the necessary embedding rate needed to refresh the keys in time T. For example, if we have a video service of QCIF images with a frame rate of 20 frames/second, and desire to refresh the keys during member departure in $T = 5$ seconds, then 25 bits must be embedded per frame. In particular, for an embeddability rate of 25 bits/frame, we note that average PSNR difference of the two test sequences is less than 1dB and therefore would introduce no noticeable distortion to the video quality. Further, in video applications that use higher-resolution video formats, such as CIF and SIF format, less distortion occurs for the same embeddability rate. Thus, for the same amount of distortion in video with a larger image size, it becomes possible to rekey larger group sizes, refresh keys faster, or increase the protection by using larger key lengths.

TABLE 6.3. Average PSNR difference.

	20 bits	40 bits	60 bits	Maximal
Foreman	0.2002(dB)	0.3054(dB)	0.4264(dB)	0.4477(dB)
Miss America	0.0720(dB)	0.1098(dB)	0.1434(dB)	0.1602(dB)

6.5   Extensions to Multilayered Services

In many application environments, the multimedia data is distributed in a multi-layered form. For example, in an HDTV broadcast, users with a normal TV receiver can still receive the current format, while other users with a HDTV receiver can receive both the normal format and the extra information needed to achieve HDTV resolution. As another example, the MPEG-4 standard allows for multiple media streams corresponding to different object planes to be composited. In either of these cases, it will be desirable for service providers to separately control access to the different layers of media. The key management schemes must therefore be considered separately, yet incorporate new key management functionalities that are not present in conventional multicast key management schemes. Specifically, it is necessary to introduce new rekeying events that allow users to subscribe or cancel membership to some layers while maintaining their membership to other layers. Hence multi-layered, or multi-object multimedia services will require additional functionality added to a multicast key management scheme.

As an example of the additional functionality needed, we use our tree-based scheme of Section 6.3.3 and consider the problem of managing keys for two levels of service corresponding to a low quality and high quality service. Extensions to more layers or objects is straight forward.

Suppose the multimedia data stream consists of two layers, which are denoted as D^l and D^h. D^l provides the low resolution service only, while high-quality service can be obtained by receiving both the base-layer D^l and the refinement-layer D^h. The GC will have two session keys $K_s^l(t)$ and $K_s^h(t)$. $K_s^l(t)$ is used to encrypt D^l and $K_s^h(t)$ is used to encrypt D^h. Similarly, each internal node in the key tree has two internal keys $K_\sigma^l(t)$ and $K_\sigma^h(t)$, where σ is the index of the nodes in the tree. Group members who want to receive the lower quality service will be assigned the low-layer session key, as well low-layer keys from the root to the leaf which stands for this member. Group members who want to receive high quality service will be assigned both the low-layer and high-layer keys. The rekeying scheme is similar to the one layer case described earlier, but requires additional functionalities since users may switch between the different levels of service.

- **Refreshing the low-quality session key:** The new session key associated with the low-quality level may be refreshed by encrypting with the root low-quality KEK $K_\epsilon^l(t)$ and transmitting the message $\alpha_s^l(t) = E_{K_\epsilon^l(t)}\left(K_s^l(t)\right)$.

- **Refreshing the high-quality session key:** The procedure for refreshing of the high-quality session key is identical to the procedure for refreshing the low-quality session key, but using $K_s^h(t)$ and $K_\epsilon^h(t)$ instead.

- **New member joins low-quality service:** A new member may desire to join the low level service. In this case, the low-quality session key and IKs must be renewed, which can be done by applying the procedure of Section 6.3.3.

- **New member joins high-quality service:** A new member may desire to join the high level service. In this case, both the low-quality and high-quality keys must be renewed. To do this, the procedure of Section 6.3.3 is applied twice, once for the low-quality keys, and once for the high-quality keys.

- **High-quality user leaves the group:** In this case, both session key $K_s^l(t-1)$ and $K_s^h(t-1)$ and corresponding IKs for both D^l and D^h have to be changed. This can be done using the algorithms in Section 6.3.3 twice.

- **Low-quality user leaves the group:** In this case, only session key $K_s^l(t-1)$ and corresponding IKs for base-layer D^l needs to be changed, which can be done using the algorithms in Section 6.3.3 once on the appropriate low-layer keys.

- **Low-quality user changes to high-quality:** In this case, the high-layer SK $K_s^h(t-1)$ as well as the high-layer IKs must be changed has to be changed to prevent the user from accessing the past high quality service. The new SK $K_s^h(t)$ and IKs keys from root to the leaf are directly given by the GC to this user during registration to the new level of service.

- **High-quality user change to low-quality:** The session key $K_s^h(t-1)$ and corresponding IKs for high-layer have to be changed to prevent this user from accessing the future high quality information. This can be done using the algorithms in Section 6.3.3 once on the high-layer internal keys.

6.6   Chapter Summary

The secure distribution of multimedia multicasts necessitates the distribution and management of keying material. In this chapter, we have examined the problem of managing keys needed to secure multimedia multicasts. We presented a new format for the rekeying messages associated with multicast key management, as well as described two modes of conveyance for transmitting the rekeying messages.

We began by discussing the fundamental problem of securely distributing information simultaneously to a group of users. This fundamental problem is at the heart of multicast key management schemes, where the information to be distributed is a new session key. We examined a simple key management scheme to motivate the importance of reducing the communication overhead associated with identifying which portion of a rekeying message is associated with each user. The communication overhead is reduced by using a homogenized message format from which every user can perform a suitable operation to extract the new keying information. We presented a homogenized message format, built using one-way functions and large integer arithmetic, that allows for each user to perform a modular operation to extract the new key information. We then examined the security of the residue-based rekeying message from an information theoretic perspective, and also showed that the residue-based rekeying message format is resistant to attacks by members of the service attempting to acquire private keying information of other members.

Typically, the information associated with rekeying is distributed via a media-independent channel. However, multimedia data allows for a media-dependent channel, such as is provided by data embedding techniques. By embedding the keying information in the multimedia content, the key updating messages associated with secure multicast key management schemes may be hidden in the data and used in conjunction with encryption to protect the data from unauthorized access. The primary advantage of using data embedding to convey rekeying messages compared to the traditional use of a media-independent channel is that data embedding hides the presence of rekeying messages from potential adversaries, thereby making it more difficult for eavesdroppers to measure information regarding membership dynamics. Further, the use of data embedding allows the application to maintain the data rate of the media without performing computationally expensive transcoding operations.

We used our proposed message form in conjunction with a data embedding technique for block-based motion compensated video compression to illustrate that the amount of time needed to update the entire network of keys is related to the amount of users in the service, key lengths used, and the embeddable channel rate. For a video service providing QCIF images with a frame rate of 20 frames/second, we observed that it was possible to refresh the keys for a group size of roughly one million users in 5 seconds when we used an embeddability rate of 25 bits/frame. The distortion introduced to the video sequence was less than 0.8dB of PSNR and was not perceptible. Finally, by adding extra functionality to multiple key trees, multicast key distribution schemes can be extended to protect multiple layers of multimedia content in an efficient manner. The additional operations needed to manage the keys for multilayered services is more complex than traditional multicast services since users may switch between different levels of service. We presented an example of a key management scheme for two levels of service, and described the necessary operations needed to allow users to drop from a high-quality service to a low-quality service, and also upgrade their service from a low-quality to a high-quality service.

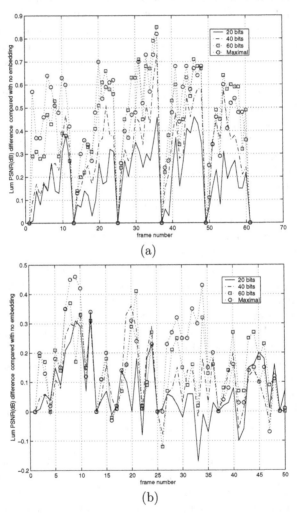

FIGURE 6.6. The peak signal-to-noise ratio (PSNR) difference of the luminance components between no embedding and the embedding scheme of Song et al. with variable embedding rate. (a) Foreman, (b) Miss America.

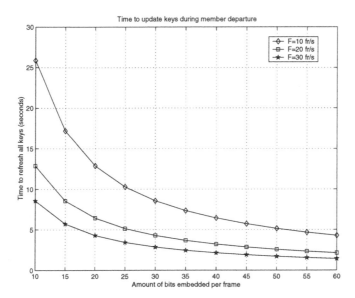

FIGURE 6.7. The time needed to refresh the entire set of keys during a member departure using the bottom-up approach with different frame rates F, and different amounts of bits embedded per frame. The group size is $n = 2^{20}$, or roughly one-million users.

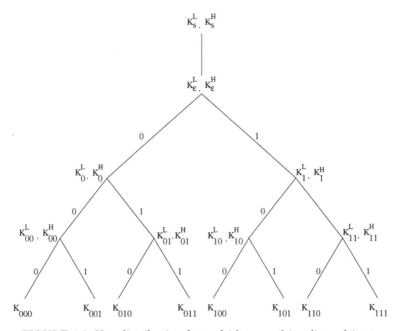

FIGURE 6.8. Key distribution for multi-layer multimedia multicast.

7
Hierarchical Access Control for Multi-Group Scenarios

Many group applications contain multiple related data streams and group members have different access privileges. These applications are prevalent in various scenarios.

- Multimedia applications distributing data in a multi-layer coding format [114]. For example, in video broadcast, users with a normal TV receiver can receive the normal format, while others with HDTV receivers can receive the normal format and the extra information needed to achieve HDTV resolution.

- Multicast programs containing several related services, such as weather, news, traffic and stock quote.

- Communications in hierarchically managed organizations where participants have various access authorization.

In these scenarios, group members subscribe to different data steams, or possibly multiple of them. In other words, the access control mechanism needs to supports multi-level access privilege. This is referred to as the *hierarchical group access control* [115,116].

Traditional key management schemes are not designed to handle key management issues associated with multiple services occurring concurrently that have correlated memberships. Although access control for individual data stream can be managed separately using existing key management schemes, this leads to inefficient use of keys and does not scale well when the number of data streams increases.

In this chapter, we formulate the hierarchical group access control problem and presents the solutions [115, 116] in both centralized and contributory environments.

7.1   Hierarchical Access Control: Problem Formulation

As the first step towards a generic solution, the hierarchical group access control problem is formulated in this section.

7.1.1   System description

Let $R = \{r_1, r_2, \cdots\}$ denote the set of *resources* in the system. In the context of group communication, one resource is one multicast data stream.

From the resource points of view, a *Data Group* (DG) is defined as all users who have access to a particular resource. It is clear that the DGs may have overlapped membership because users may subscribe to multiple resources.

From access control points of view, a *Service Groups* (SG) is defined as a set of users who can access the exactly same set of resources. SGs do not have overlapped membership. In this chapter, the DGs are denoted by $\{D_1, D_2, \cdots, D_M\}$, where M is the total number resources, and the SGs are denoted by $\{S_1, S_2, \cdots, S_I\}$, where I is the total number SGs. It is easy to prove that $I \leq 2^M - 1$.

The access relationship between the resources and the SGs can be described by a *capability list*. Here are two examples illustrating typical access relationship in group communication.

Example 1. Multimedia applications that distribute data in a multi-layered format [114].

- *Resources*: {base layer (r_1), enhancement layer 1 (r_2), enhancement layer 2 (r_3)}.

- *Service Groups*: {users subscribing basic quality (S_1), users subscribing moderate quality (S_2), users subscribing high quality (S_3)}.

- *Capability lists*:S_1 access $\{r_1\}$; S_2 access $\{r_1, r_2\}$; S_3 access $\{r_1, r_2, r_3\}$.

Example 2. Multicast programs containing several related services.

- *Resources*: {news (r_1), stock quote (r_2), traffic/weather (r_3)}.

- *Service Groups*: users can subscribe to any combination of the resources. Thus, there are total 7 SGs, denoted by $S_1, S_2, \cdots, S_7$.

- *Capability lists*: S_1 access $\{r_1\}$; S_2 access $\{r_2\}$; S_3 access $\{r_3\}$; S_4 access $\{r_1, r_2\}$; S_5 access $\{r_1, r_3\}$; S_6 access $\{r_2, r_3\}$; S_7 access $\{r_1, r_2, r_3\}$.

Besides the capability list, *access matrix* is also used to describe access relationship. In particular, the element on the ith row and m^{th} column of the access matrix, denoted by $a_{i,m}$, is

$$a_{i,m} = \begin{cases} 1, & \text{if SG } S_i \text{ can access resource } r_m \\ 0, & \text{otherwise} \end{cases},$$

where $i = 1, \cdots, I$ and $m = 1, \cdots, M$.

Based on those definitions, the group size of SGs and DGs must satisfy:

$$n(D_m) = \sum_{i=1}^{I} a_{i,m} \cdot n(S_i), \tag{7.1}$$

where $n(S_i)$ is the number of users in SG S_i and $n(D_m)$ is the number of users in DG D_m.

7.1.2 Security requirements

In the applications containing multiple multicast sessions, users not only join or leave service, as addressed in the single multicast session scenario, but also may switch between the SGs by subscribing or dropping data streams. Thus, the security requirements are more complicated than these for a single multicast session.

Let the notation $S_i \rightarrow S_j$ represent a user switching from SG S_i to SG S_j. To simplify future notations, S_0 is defined as a virtual service group containing users who cannot access any resources. Thus, $S_0 \rightarrow S_i$ represents a user joining SG S_i, and $S_i \rightarrow S_0$ represents a user leaving the group communication from SG S_i.

Similar to the single session access control problem addressed by the traditional key management schemes [78], the hierarchical group access control should guarantee the following security requirements.

- The users in the SG S_i have and only have access to resources $\{r_m, \forall\, m : a_{i,m} = 1\}$.

- When a user $S_i \rightarrow S_j$,

    - This user cannot access the future content of the resources $\{r_m, \forall\, m : a_{i,m} = 1 \text{ and } a_{j,m} = 0\}$. This property is referred to as the forward secrecy [66].

    - This user cannot access the previous content of the resources $\{r_m, \forall\, m : a_{i,m} = 0 \text{ and } a_{j,m} = 1\}$. This property is referred to as the backward secrecy [66].

7.1.3 Data encryption and hierarchical key management

To formulate the hierarchical access problem, the ways of encrypting multiple data streams need to be clarified first. In the hierarchical access control scenario, there are two ways to encrypt and distribute multicast data. In the first method, resources are encrypted using separate keys, which are called *Data Group Keys*. The data group key used to encrypt resource r_m, denoted by K_m^D, is shared among the users in DG D_m. In this case, each resource is distributed in a single multicast session, and the users may subscribe to one or several multicast sessions according to their access privilege. The task of key management is to securely update and distribute $\{K_m^D, \forall m : a_{i,m} = 1\}$ to the users in S_i, where $i = 1, 2, \cdots, I$.

In the second method, the users in each SG share a secrete key called the *Service Group Key* and the multicast sessions are formed based on SGs. In particular, the users in S_i share the service group key K_i^S and form one multicast session. In this multicast session, the resources $\{r_m, \forall m : a_{i,m} = 1\}$ are encrypted by K_i^S and transmitted to the users in S_i. In this case, one resource may be distributed in several multicast sessions while being encrypted by different service group keys. The task of key management is to securely distribute and update K_i^S for the users in SG S_i.

To illustrate these two methods, let's re-visit Example 1 described in section 7.1.1.

In the first method, data are transmitted in three multicast sessions. The first session contains all users, and distributes resource r_1 encrypted by K_1^D. The second session contains users in S_2 and S_3, and distributes resource r_2 encrypted by K_2^D. The third session contains users in S_3, and distribute resource r_3 encrypted by K_3^D. The communication overhead of a multicast session can be described as $DR(r_i)G(x)$, where $DR(r_i)$ denotes the data rate of resource r_i, and $G(x)$ is the cost of sending unit data to x users through multicast. The first method has communication overhead as $CommCost_{method_1} = DR(r1)G(n(S_1) + n(S_2) + n(S_3)) + DR(r2)G(n(S_2) + n(S_3)) + DR(r3)G(n(S_3))$.

When using the second method, there are three multicast sessions also. The first session contains users in S_1, and distributes resource r_1 encrypted by K_1^S. The second session contains users in S_2, and distributes resource r_1 and r_2 encrypted by K_2^S. The third session contains users in S_3, and distribute all three resources encrypted by K_3^S. The communication overhead is $CommCost_{method_2} = DR(r1)G(n(S_1)) + (DR(r1) + DR(r2))G(n(S_2)) + (DR(r1) + DR(r2) + DR(r3))G(n(S_3))$. Using the fact the $G(x+y) < G(x) + G(y)$ in multicast communications, it can be seen that $CommCost_{method_2} > CommCost_{method_1}$.

On the other hand, users in the second method only subscribe to one multicast session. Thus, the task of key management for the second method can be solved by applying traditional key management for each SG separately.

We suggest to adopt the first encryption method because of its low data communication overhead. When the first encryption method is used, in order to guarantee forward and backward secrecy, when a user switches from SG S_i to S_j, the proposed key management scheme should

- update $\{K_m^D, \forall m : a_{i,m} = 0 \text{ and } a_{j,m} = 1\}$, such that this user cannot access the previous communication in corresponding DGs;

- and update $\{K_m^D, \forall m : a_{i,m} = 1 \text{ and } a_{j,m} = 0\}$, such that this user cannot access the future communication in corresponding DGs.

7.2 Centralized Multi-group Key Management Scheme

Hierarchical group access control can be achieved in either centralized or contributory manner. While the contributory solution will be discussed in Section 7.5, this section and the following two sections will be dedicated to the centralized schemes.

7.2.1 Independent key trees for hierarchical access control

To reduce the communication, computation and storage overhead, tree structure is widely used in centralized key management schemes to maintain the keying material and coordinate the key generation [7, 8, 10, 79, 80, 117, 118] (see Section 5.1).

When using tree-based schemes to achieve hierarchical group access control, a separate key tree must be constructed for each DG, with the root being the data group key and the leaves being the users in this DG. This approach is referred to as the *Independent-tree* key management scheme, and is illustrated in Figure 7.1. This scheme does not exploit the relationship among the subscribers and makes inefficient use of keys because of the overlapped DG membership. As an extreme example, if a user who subscribes all data streams leaves the service, key updating has to take place on all key trees.

7.2.2 Multi-group key management scheme

We develop a *multi-group* key management scheme that employs one integrated key graph accommodating key materials of all users. This key graph consists of several subtrees, and is constructed in three steps.

Step1: For each SG S_i, construct a subtree having the leaf nodes as the private keys of users in S_i and the root node as the service group key K_i^S. These subtrees are referred to as the *SG-subtrees*.

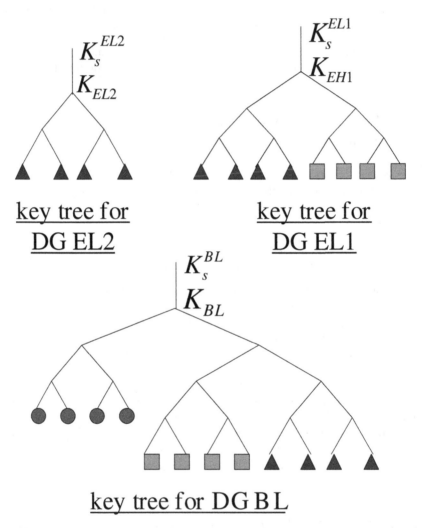

FIGURE 7.1. Independent-tree key management scheme for layered coded multimedia service

Step 1:

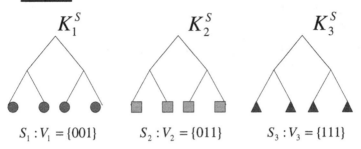

$$S_1 : V_1 = \{001\} \qquad S_2 : V_2 = \{011\} \qquad S_3 : V_3 = \{111\}$$

Step 2:

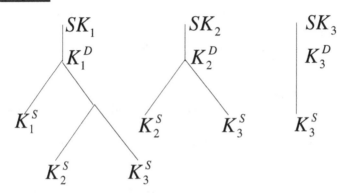

Step 3:

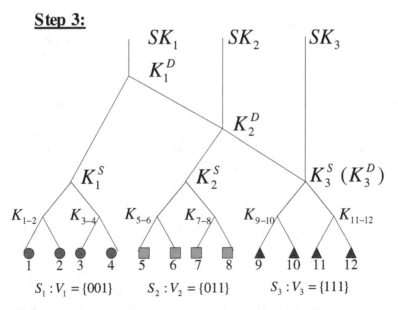

$$S_1 : V_1 = \{001\} \qquad S_2 : V_2 = \{011\} \qquad S_3 : V_3 = \{111\}$$

FIGURE 7.2. Multi-group key management graph construction

Step2: For each DG D_m, construct a subtree whose root is the DG key K_m^D and whose leaves are $\{K_i^S, \forall i : a_{i,m} = 1\}$. These subtrees are referred to as the *DG-subtrees*.

Step3: Generate the key graph by connecting the leaves of the DG-subtrees and roots of SG-subtrees.

---

for $i = 1 : I$ **do**
    Construct a tree, called SG-subtree of S_i, with $n(S_i)$ leaf nodes.;
    Assign users in S_i to leaf nodes.;
end
for $m = 1 : M$ **do**
    Construct a tree, called DG-subtree of D_m, with $\sum_i a_{i,m}$ leaf nodes.;
    Assign key $\{K_i^S, \forall i : a_{i,m} = 1\}$ to leaf nodes, and K_m^D to the root node.;
end
for $i = 1 : I$ **do**
    Search all leaf nodes of DG subtrees and find these are associated with K_i^S;
    Merge these nodes and the root of the SG-subtree of S_i into one node.;
end

Algorithm 13: Integrated Key Graph Generation

This 3-step procedure is formally described in Procedure 13 and illustrated in Figure 7.2 for the service containing 3 layers and 4 users in each SG. In the first and the second step, there is no constraint on the tree structures that can be used for the SG- and DG-subtrees. Binary subtrees are used to demonstrate the performance of the multi-group key management. There is a lot of flexibility in subtree design. For example, when considering the heterogeneity among SGs, the DG-subtrees can be designed as unbalanced trees. In the third step, some duplicated structures may appear on DG-subtrees. In Figure 7.2, the DG-subtrees of D_2 and D_1 have the same structures that connect K_2^S and K_3^S. Duplicated structures can be merged. In this example, the parent node of K_2^S and K_3^S on DG-subtree of D_2 are merged with K_2^D. This merging operation can further reduce the number of keys on the key graph, but the effect of merging is very small especially when the group size is large. Therefore, the performance analysis in Section 7.3 does not consider the positive effect of merging, and thus provides the performance upper bound.

This multi-group key graph can also be interpreted as M overlapped key trees, each of which has K_m^D as the root and the users in DG D_m as the leaves. Obviously, these M key trees can be used in the independent-tree

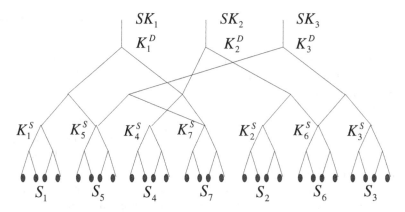

FIGURE 7.3. An integrated key graph for multiple service scenario (Example 2, Section II-A)

scheme. This reveals the fact that the multi-group key graph removes the "redundancy" presented in the independent-tree scheme. Therefore, it can reduce overhead.

As another example, Figure 7.3 shows a multi-group key graph for the multiple service scenario described in Example 2 in Section 7.1. It is noted that neither the design of the DG-subtrees nor the merging operation is unique. Although the graph between the DG keys and the SG keys can be optimized to minimize the number of keys on the graph, this optimization introduces little gain but high computational complexity. Therefore, the proposed scheme does not specify how to merge the DG-subtrees. In most cases, it is not necessary to merge the DG-subtrees.

As defined in [8], *keyset* refers to the set of keys associated with an edge node on the key graph and possessed by the user located at this edge node. In our key graph, the keyset of a user in SG S_i is the keys on the pathes from itself to the roots of the DG-subtrees of D_m for $\{m : a_{i,m} = 1\}$. It is noted that the keyset of users in S_0 is an empty set.

Besides user join and departure, the rekey algorithm in the multi-group key management scheme must address users' relocation on the key graph. The rekey algorithm for $S_i \rightarrow S_j$, which includes the cases for user join, departure, and switching, is described as follows.

When a user switches from S_i to S_j, the switching user is moved from the SG-subtree of S_i to a new location on the SG-subtree of S_j. Let ϕ_i denote the keyset associated with the user's previous position, and ϕ_j denote the keyset associated with the user's new position. Then,

- the KDC updates the keys in $\overline{\phi_i} \cap \phi_j$ using one-way functions, similar to the procedure for user join described in Section 7.2.1,

- and, the KDC generates new versions of the keys in $\phi_i \cap \overline{\phi_j}$ and distributes these new keys encrypted by their children node keys from bottom to up, similar to the procedure for user departure described in Section 7.2.1.

This rekey algorithm is illustrated based on the sample key tree shown in Figure 4. Let user 8 switches from SG S_2 to S_1. The key tree is updated as shown in Figure 7.4. On the SG-subtree of S_1, the leaf node associated with user 4 is split in order to accommodate user 8. Then, user 4 and 8 share a new KEK, denoted by K_{4-8}. On the SG-subtree of S_2, user 7 is moved up and occupy the node that is previously associated with K_{7-8}. In this case, ϕ_2 is $\{K_{7-8}, K_2^S, K_2^D, SK_2, K_1^D, SK_1\}$ and ϕ_1 is $\{K_{4-8}, K_{3-4}, K_1^S, K_1^D, SK_1\}$.

The KDC generates the new keys, K_{3-4}^{new} and $K_1^{S,new}$, from the old keys using a one-way function, and increases the revision numbers of those new keys. Thus, the user 1,2,3,4 will know about the key change when the data packet indicating the increase of the revision numbers first arrives, and compute the new keys using the same one-way function. No rekeying messages are necessary for K_{3-4}^{new} and $K_1^{S,new}$.

Then, the KDC generates new keys, K_{4-8}^{new}, $K_2^{S,new}$, $K_2^{D,new}$, and SK_2^{new}, and distributes them through a set of rekeying messages as:

$$\{K_{4-8}^{new}\}_{u_8}, \{K_{4-8}^{new}\}_{u_4}, \{K_2^{S,new}\}_{K_{5-6}}, \{K_2^{S,new}\}_{u_7}$$
$$\{K_2^{D,new}\}_{K_2^{S,new}}, \{K_2^{D,new}\}_{K_3^S}, \{SK_2^{new}\}_{K_2^{D,new}}.$$

In this case, the rekeying message size is 7.

It is noted that $\overline{\phi_i} \cap \phi_j$ may contain the new KEKs that are created for accommodating the switching user. These new KEKs are encrypted by users' private keys and distributed through rekeying messages. In addition, $\phi_i \cap \overline{\phi_j}$ may contain KEKs that do not exist any more after the relocation of the switching user. Obviously, these keys are discarded.

7.3   Performance Measures and Analysis

Communication, computation and storage overhead associated with key updating are major performance criteria for key management schemes [7, 8, 78]. To measure the performance of hierarchical access control schemes, the performance metrics are defined as follows.

- *Storage overhead at the KDC*: denoted by R_{KDC} and defined as the expected number of keys stored at the KDC.

- *Rekey overhead at the KDC*: denoted by M_{KDC} and defined as the expected amount of rekeying messages transmitted by the KDC.

- *Storage overhead of users*: denoted by $R_{u \in S_i}$ and defined as the expected number of keys stored by the users in SG S_i.

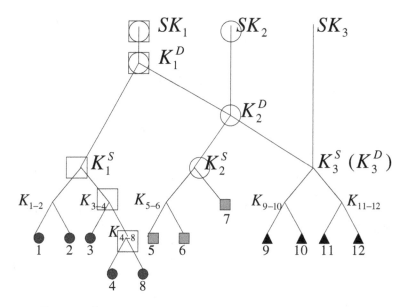

ϕ_1 : keyset of the previous position

ϕ_2 : keyset of the new position

FIGURE 7.4. User relocation on the key graph

- *Rekey overhead of users*: denoted by $M_{u \in S_i}$ and defined as the expected amount of rekeying messages received by the users in SG S_i.

Here, R_{KDC} and $R_{u \in S_i}$ describe the storage overhead, whereas M_{KDC} and $M_{u \in S_i}$ reflect the usage of communication and computation resources. For example, given the group size and network topology, M_{KDC} is proportional to the total amount of key management data forwarded on the network.

For the interested readers, many intermediate results and derivations in section 7.3.1 and section 7.3.2 can be applied to analyzing other types of tree-based key management schemes. Readers can also jump to (7.11) and (7.14) for asymptotic storage overhead results, and (7.19) for asymptotic communication and computation overhead results.

7.3.1 Storage overhead

The storage overhead of a single key tree is considered first. Similar to most key management schemes [7, 8, 10, 78, 80], the key tree investigated here is fully loaded and maintained as balanced as possible by putting the joining users on the shortest branches.

Let $f_d(n)$ denote the length of the branches and $r_d(n)$ denote the total number of keys on the key tree when the key tree has degree d and accommodates n users. Since the key tree is balanced, the length of the branches can be either $\lfloor \log_d n \rfloor$ or $\lfloor \log_d n \rfloor + 1$. In this chapter, we use L_0 to represent the length of the shortest branch, i.e. $L_0 = \lfloor \log_d n \rfloor$, and $f_d(n)$ is either L_0 or $L_0 + 1$. Additionally,

- the number of users who are on the branches with length L_0 is $d^{L_0} - \lceil \frac{n - d^{L_0}}{d-1} \rceil$,

- and, the number of users who are on the branches with length $L_0 + 1$ is $n - d^{L_0} + \lceil \frac{n - d^{L_0}}{d-1} \rceil$.

On the key tree, there are 1 session key, n users' private keys, $(1 + d + d^2 + \cdots + d^{L_0 - 1})$ key encrypted keys in the upper $L_0 - 1$ levels, and $\lceil \frac{n - d^{L_0}}{d-1} \rceil$ key encrypted keys at the L_0 level. Therefore, the total number of keys on this key tree is calculated as

$$r_d(n) = n + 1 + \frac{d^{L_0} - 1}{d - 1} + \lceil \frac{n - d^{L_0}}{d - 1} \rceil. \tag{7.2}$$

Using the fact that $\frac{n - d^{L_0}}{d-1} \leq \lceil \frac{n - d^{L_0}}{d-1} \rceil < \frac{n - d^{L_0}}{d-1} + 1$, from (7.2), one can derive

$$\frac{dE[n] - 1}{d - 1} + 1 \leq E[r_d(n)] < \frac{dE[n] - 1}{d - 1} + 2, \tag{7.3}$$

where the expectation, $E[.]$, is taken over the distribution of $n(D_m)$ and the length of the branches on the key trees. The left-hand-side equality is achieved when $\log_d(n)$ is an integer.

Next, we calculate $E[f_d(n)]$. Since $\lfloor \log_d n \rfloor \leq \log_d n$ and $f_d(n) < L_0 + 1$, we have $E[f_d(n)] < E[\lfloor \log_d n \rfloor + 1] \leq E[\log_d n] + 1$. Since $\log_d(n)$ is a concave function, $E[\log_d n] < \log E[n]$. We can see that

$$E[f_d(n)] \leq \log_d E[n] + 1. \tag{7.4}$$

With (7.3) representing the total number of keys and (7.4) representing the number of keys on a branch, the storage overhead can be analyzed. When using the separate key trees, the KDC stores all keys on total M key trees, and users in S_i store subsets of keys on the key trees that are associated with D_m, for $\{m : a_{i,m} = 1\}$. Thus,

$$R^{ind}_{KDC} = \sum_{m=1}^{M} E[r_d(n(D_m))], \tag{7.5}$$

$$R^{ind}_{u \in S_i} = \sum_{m=1}^{M} a_{i,m} \left(E[f_d(n(D_m))] + 1 \right). \tag{7.6}$$

In the multi-group key management scheme, the DG-subtree of D_m has $c_m = \sum_i a_{i,m}$ leaf nodes. Before removing the redundancy on DG-subtrees, there are in total $\sum_{m=1}^{M} r_d(c_m)$ keys on DG-subtrees. Also, the total number of keys on the SG-subtrees is $\sum_{i=1}^{I} r_d(n(S_i))$. Merging duplicated structures on DG-subtrees only reduces the number of keys on the key graph. Therefore, the storage overhead at the KDC is

$$R_{KDC}^{mg} \leq \sum_{i=1}^{I} E[r_d(n(S_i))] + \sum_{m=1}^{M} E[r_d(c_m)]. \tag{7.7}$$

A user in the SG S_i stores $f_d(n(S_i))$ keys on the SG-subtree and up to $\sum_{m=1}^{M} a_{i,m}(f_d(c_m)+1)$ keys on the DG-subtrees. Therefore, the users' storage overhead of the multi-group scheme is:

$$R_{u \in S_i}^{mg} \leq E[f_d(n(S_i))] + \sum_{m=1}^{M} a_{i,m}(E[f_d(c_m)] + 1). \tag{7.8}$$

Next, the storage overhead is analyzed in the applications containing multiple layers, as described in Example 1 in Section 7.1.1. This is a special application scenario for the multi-group key management schemes. In this scenario, $a_{i,m} = 1$ for $m \leq i$ and $a_{i,m} = 0$ for $m > i$. It is assumed that each layer contains the same amount of users, denoted by $n(S_i) = n_0$. Thus, $n(D_m) = (M - m + 1)n_0$. Using (7.6) and (7.8), the users' storage overhead is calculated as:

$$R_{u \in S_i}^{ind} = \sum_{m=1}^{i} \left(E[f_d((M - m + 1) \cdot n_0)] + 1 \right), \tag{7.9}$$

$$R_{u \in S_i}^{mg} \leq E[f_d(n_0)] + \sum_{m=1}^{i} \left(E[f_d(M - m + 1)] + 1 \right). \tag{7.10}$$

When the group size is large, i.e. $n_0 \to \infty$, (7.4)(7.9) and (7.10) lead to

$$R_{u \in S_i}^{ind} \sim O(i \cdot \log(n_0)), \quad R_{u \in S_i}^{mg} \sim O(\log(n_0)). \tag{7.11}$$

Based on (7.5) and (7.7), the storage overhead at the KDC is calculated as:

$$R_{KDC}^{ind} = \sum_{m=1}^{M} E[r_d(m \cdot n_0)], \tag{7.12}$$

$$R_{KDC}^{mg} \leq M \cdot E[r_d(n_0)] + \sum_{m=1}^{M} E[r_d(m)]. \tag{7.13}$$

From (7.3), it is seen that $\lim_{n \to \infty} r_d(n) = \frac{d}{d-1}n$. Then, from (7.12) and (7.13), one can derive

$$R_{KDC}^{ind} \sim O(\frac{d}{d-1} \frac{M(M+1)}{2} n_0);$$

$$R_{KDC}^{mg} \sim O(\frac{d}{d-1} M \cdot n_0). \tag{7.14}$$

From above results, two observations are made. First, by using the integrated key graph instead of the separate key trees, the multi-group key management scheme reduces the storage overhead of both the KDC and the users. Second, as indicated in (7.14), the storage advantage of the multi-group scheme becomes larger when the system contains more SGs, i.e. requiring more levels of access control. When the number of layers (M) increases, the multi-group scheme scales better than the independent-tree scheme.

7.3.2 Rekey overhead

Let $C_{i,j}$ denote the amount of rekeying messages transmitted by the KDC when one user switches from S_i to S_j. It is noted that the rekey overhead, M_{KDC} and $M_{u \in S_i}$, can be calculated from $C_{i,j}$, as long as the users' statistical joining/leaving/switching model is given.

Switching from S_i to S_j is equivalent to adding the subscription to $\{D_m, \forall m : a_{i,m} = 0 \text{ and } a_{j,m} = 1\}$ and dropping the subscription to $\{D_m, \forall m : a_{i,m} = 1 \text{ and } a_{j,m} = 0\}$. When using the tree-based key management schemes, the rekeying message size is:

$$C_{ij}^{ind} = \sum_{m=1}^{M} \max(a_{i,m} - a_{j,m}, 0) \cdot (d \cdot f_d(n(D_m))). \qquad (7.15)$$

A brief explanation of (7.15) is as follows. The term $(\max(a_{i,m} - a_{j,m}, 0))$ equals to 1 only when $a_{i,m} = 1$ and $a_{j,m} = 0$. When this term equals to 1, keys need to be updated on the key tree of DG D_m. Recall that d messages are needed to update one KEK and the length of the branch is $f_d(n(D_m))$. Therefore, $d \cdot f_d(n(D_m))$ rekeying messages are necessary.

In the multi-group key management scheme, when a user switches from S_i to S_j and $i \neq j$, the amount of messages needed to update keys on the SG-subtree of S_i is up to $(d \cdot f_d(n(S_i))) - 1$. The amount of messages needed to convey the KEK created for accommodating the switching/join user on the SG-subtree of S_j is no more than 2. If this user drops the subscription of the DG D_m, i.e. $(\max(a_{i,m} - a_{j,m}, 0)) = 1$, the amount of rekeying messages that update keys on the DG-subtree of D_m is up to $(d \cdot f_d(c_m) + 1)$. If this user remains the subscription of the DG D_m, i.e. $a_{i,m} = a_{j,m} = 1$, up to $(d \cdot f_d(c_m))$ rekeying messages are needed to update keys on the DG-subtree of D_m. Therefore, when using the multi-group scheme and $i \neq j$, the C_{ij} value is

$$C_{ij}^{mg} \leq \sum_{m=1}^{M} (\max(a_{i,m} - a_{j,m}, 0) \cdot (d \cdot f_d(c_m) + 1)$$
$$+ a_{i,m}a_{j,m}d \cdot f_d(c_m)) + d \cdot f_d(n(S_i)) + 1. \qquad (7.16)$$

Similar to that in Section 7.3.1, the rekey overhead in a multi-layer scenario with $n(S_i) = n_0$ is analyzed. In this application scenario, the rekeying message size for one user departure, i.e. $S_j \to S_0$, is computed from (7.15) and (7.16) as:

$$C_{0j}^{ind} = \sum_{m=1}^{j} d \cdot E[f_d((M - m + 1)n_0)], \qquad (7.17)$$

$$C_{0j}^{mg} \leq d \cdot E[f_d(n_0)] + 1$$
$$+ \sum_{m=1}^{j} (d \cdot E[f_d(M - m + 1)] + 1). \qquad (7.18)$$

When $n_0 \to \infty$, it is straightforward to derive

$$C_{0j}^{ind} \sim O(j \cdot d \cdot \log(n_0)), \ C_{0j}^{mg} \sim O(d \cdot \log(n_0)). \qquad (7.19)$$

The comprehensive comparison between the multi-group scheme and the independent-tree scheme will be presented later in Section 7.4.

7.4 Simulations and Performance Comparison

In this section, the multi-group key management scheme and independent-tree scheme are compared in various application scenarios.

7.4.1 Statistical dynamic membership model

In [90] [72], it has been shown that the users' arrival process and membership duration of MBone multicast sessions can be modelled by Poisson and exponential distribution respectively, in a short period of time. In this chapter, we use this Poisson arrival and exponential distribution duration model. In addition, it is assumed that when a user switches between SGs, the SG that he switches to depends only on his current SG.

As a result of these assumptions, the users' statistical behavior can be described by an embedded Markov chain [92]. Particularly, there are a total of $I + 1$ states, denoted by $\tilde{S}_i$, $i = 0, \cdots, I$. When a user is in the SG S_i, he is in the state $\tilde{S}_i$. After a user enters state $\tilde{S}_i$, i.e. subscribes or switches to SG S_i, this user stays at state $\tilde{S}_i$ for time T_i, which is governed by an exponential random variable. When time is up, the user moves to state $\tilde{S}_j$. The selection of $\tilde{S}_j$ only depends on the current state $\tilde{S}_i$ and is not related to previous states.

In practice, it is usually not necessary to update keys immediately after membership changes. Many applications allow the join/departure users receive limited previous/future communications [119]. For example, in video streaming applications, a joining user may receive a complete group-of-picture (GOP) [114] although partial of this GOP already been transmitted before his subscription. Those situations prefer *batch rekeying* [119]

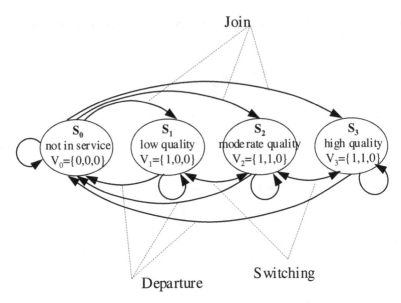

FIGURE 7.5. Discrete Markov chain model for multi-layer applications.

that postpones key updating such that the rekeying overhead is reduced by adding or removing several users altogether.

Usually, batch rekeying is implemented as updating keys periodically. The time between key updates is fixed and denoted by B_t. For the users who join/leave/switch SGs in the time interval $((k-1)B_t, kB_t]$, the key updating will take place at time kB_t, where k is a positive integer. From the key updating points of view, with batch rekeying, one can prove that the previous continuous Markov model can be simplified as a discrete Markov chain model [92], as illustrated in Figure 7.5. In this model,

- The transition matrix is denoted by $P = [p_{ij}]_{(I+1) \times (I+1)}$, where p_{ij} is the probability that one user moves from SG S_i to S_j in the time interval $(kB_t, (k+1)B_t]$ given that this user is in S_i at time kB_t.

- The n-step transition probability matrix is denoted by $P(n)$, and obviously, $P(n) = P^N$. The element at the ith row and j^{th} column of $P(n)$ is denoted by $p_{ij}(n)$.

- The stationary state probability is a 1-by-$(I+1)$ vector, denoted by $\pi = [\pi_0, \pi_1, \cdots, \pi_I]$.

In practice, most group applications have the following properties.

- $p(n)_{0j} \neq 0$ for some positive finite n and for any j because users should be able to subscribe to any SGs.

- $p(n)_{i0} \neq 0$ for some positive finite n and for any i because users should be able to leave from any SGs.

- $p_{ii} > 0$ because users can always stay in his current SG.

- The mean recurrence time [92] of the state $\tilde{S}_0$ is finite because the expected time that a user stays in the group communication is finite.

Because of these properties, this Markov chain is irreducible, aperiodic and positive recurrent. As a result, the stationary state probability mass function (pmf) exists [92] and is the unique solution of

$$\pi P = \pi, \text{ and } \sum_i \pi_i = 1 . \qquad (7.20)$$

7.4.2 Performance with different group size

We first study the applications containing multiple layers (see Example 1 in Section 7.1.1) where users in SG S_i can access DG $D_1, D_2, \cdots, D_i$. In the simulation, the transition matrix is chosen as follows.

- Users join different SGs with the same probability, i.e. $P_{0j} = \alpha, \forall j > 0$.

- Users leave different SGs with the same probability, i.e. $P_{i0} = \beta, \forall i > 0$.

- While a user is in the service, he adds/drops only one DG at a time, i.e. $P_{i,j} = 0, \forall i, j > 0$ and $|i - j| > 1$. Also, users switch between SGs with the same probability, i.e. $P_{i,j} = \gamma, \forall i, j > 0$ and $|i - j| = 1$.

Thus, the transition matrix is described by only three variables. For example, the multi-layer service with $M = 3$ has the transition matrix as:

$$P = \begin{bmatrix} 1 - 3\alpha & \alpha & \alpha & \alpha \\ \beta & 1 - \beta - \gamma & \gamma & 0 \\ \beta & \gamma & 1 - \beta - 2\gamma & \gamma \\ \beta & 0 & \gamma & 1 - \beta - \gamma \end{bmatrix}$$

In all simulations, batch rekeying is applied and the key trees are binary. The initial state is chosen as the stationary state, i.e. S_i contains $N_0 \pi_i$ users at the beginning of the service.

In Figure 7.6, 7.7, 7.8 and 7.9, the multi-group scheme and the independent-tree scheme are compared for varying group size, N_0. The results are averaged over 300 realizations, and the number of layers is 4. In those simulations, $\alpha = 0.005$, $\beta = 0.01$, and $\gamma = 0.001$.

Figure 7.6 shows that the storage overhead at the KDC, R_{KDC}, increases linearly with the group size. This result can be verified by (7.3)(7.5) and (7.7). In the case when $M = 4$, the multi-group scheme reduces R_{KDC} by more than 50%.

Figure 7.7 shows that the users' storage overhead, $R_{u \in S_i}$, increases linearly with the logarithm of the group size. This can be verified by (7.9)

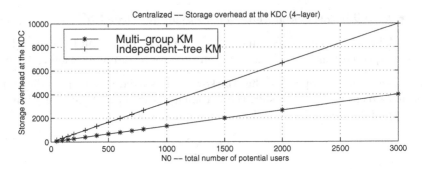

FIGURE 7.6. Storage overhead at the KDC

and (7.10). The users who subscribe only one layer have the similar storage overhead in both schemes. For the uses who subscribe multiple layers, the multi-group scheme results in less storage overhead than the independent-tree scheme.

The KDC's rekeying overhead, R_{KDC} and the users' rekey overhead, $R_{u \in S_i}$ are shown in Figure 7.8 and 7.9, respectively. In both cases, the multi-group scheme reduces the rekey overhead by more than 50%.

7.4.3   Scalability

Next, we change the number of layers (M) while maintaining roughly the same number of users in the service by choosing the join probability α as $0.02/M$. The values of β and γ are the same as those in Section 7.4.2.

Figure 7.10(a) and Figure 7.11(a) show the storage and rekey overhead at the KDC, respectively. When M increases, the storage and rekey overhead of the multi-group scheme do not change much, while the overhead of the independent-tree scheme increases linearly with M. It is clear that the multi-group scheme scales better when M increase. By removing the redundancy in DG membership, the scale of the key graph mainly depends on the group size, not the number of layers or services. On the other hand, by constructing M separate key trees, the independent-tree scheme requires larger storage and rekey overhead when M increases even when N_0 is fixed.

Figure 7.10(b) shows that the ratio between R_{KDC}^{ind} and R_{KDC}^{mg} increases linearly with M, which agrees with (7.14). Similarly, the ratio between M_{KDC}^{ind} and M_{KDC}^{mg} increases linearly with M, as shown in Figure 7.11(b).

7.4.4   Performance with different transition probability

In the previous experiments, γ is set to be 0.1β. This means that the users are more likely to leave the service than to switch SGs. Figure 7.12 shows the rekey overhead with different values of γ. Remember that γ describes the probability of user switching between SGs. In this simulation, $M = 4$,

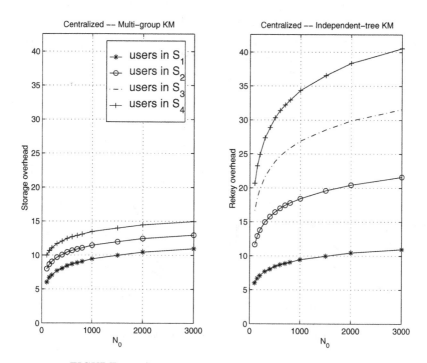

FIGURE 7.7. Storage overhead at the users in each SG

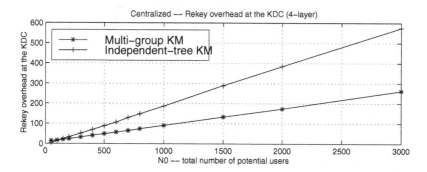

FIGURE 7.8. Rekey overhead at the KDC

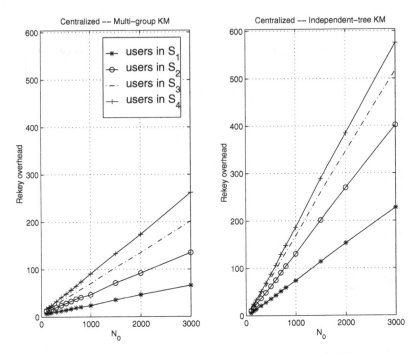

FIGURE 7.9. Rekey overhead at the users in each SG

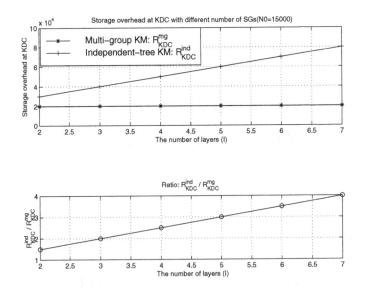

FIGURE 7.10. Storage overhead at the KDC with different number of SGs

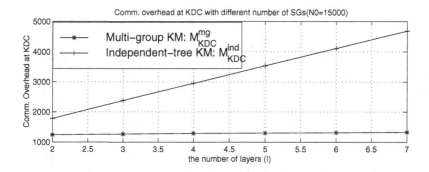

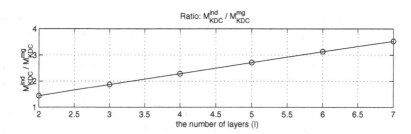

FIGURE 7.11. Rekey overhead at the KDC with different number of SGs

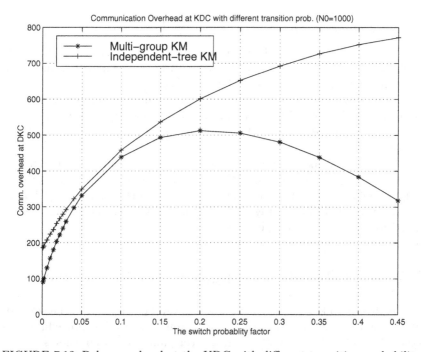

FIGURE 7.12. Rekey overhead at the KDC with different transition probability

$N_0 = 1000$, and the values of α and β are the same as those in the previous experiments.

When γ is very small, the multi-group scheme reduces the rekey overhead by about 50%, as we have shown in the previous simulations. When γ is less than 2β, the advantage of the multi-group scheme decreases with the increase of γ. This is because the multi-group scheme introduces larger rekey overhead when users switch SGs by simply subscribing more DGs. To see this, let a user move from SG S_1 to SG S_2. When using the independent-tree scheme, this user only needs to be added to the key tree associated with the DG D_2 and no rekeying messages are necessary. When using the multi-group scheme, we need to update keys on the SG-subtree of S_1 and the DG-subtree of D_1. Therefore, the performance gain reduces when more users tend to switch SGs.

When γ continues to increase, however, the rekey overhead of the multi-group scheme decreases. Particularly, when $\gamma = 0.45$, which describes the scenario where users are much more likely to switch SGs than to stay in the current SG or leave the service, the performance gain of the multi-group scheme is about 50% again. This phenomena is due to the fact that the size of the SG-subtree is greatly reduced when a significant potion of users are switching away from this SG. In this case, removing a large potion of users from the key tree using batch rekeying requires less rekeying messages than just removing several users.

7.4.5 Simulation of multi-service applications

The next experiment is for the multi-service scenario illustrated in Example 2 (Section 7.1.1), which contains 3 DGs and 7 SGs. The users can subscribe any combination of DGs and switch to any SGs. Here, the transition matrix is 8 by 8, with $P_{j0} = 0.01, \forall j > 0$ and $P_{i,j} = 0.00017, \forall i, j > 0$ and $i \neq j$. N_0 is fixed to be 1500. The values of $P_{0i}, \forall i > 0$, are adjusted such that the SGs contain varying number of users while $(\sum_{i=1}^{i} P_{0i})$ is maintained to be the same.

The horizontal axis in Figure 7.13 is the ratio between the number of users subscribing more than one DGs and the number of users subscribing only one DG. Larger is this ratio, more overlap is in DG membership. Figure 7.13 shows that the advantages of the multi-group scheme is larger when more users subscribe multiple DGs.

7.5 Contributory Multi-group Key Management

The multi-group key management schemes can be extended to the contributory environment by using the same graph construction procedure presented in Section 7.2.2. Similar as in the centralized environments, separate key trees for each DG must be constructed when using existing tree-based

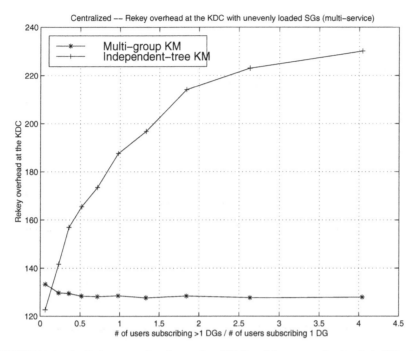

FIGURE 7.13. Rekey overhead at the KDC with unevenly loaded SGs in multi-service applications

contributory schemes [61,66,67], and the multi-group contributory schemes maintains one integrated key graph for all users.

The key establishment protocols are straightforward extensions from the existing protocols in tree-based contributory schemes [61, 66, 67]. When users join/leave/switch, the set of keys that need to be recalculated is the same as that need to be updated in the protocols presented in Section 7.2.2. The new keys are recalculated by applying the DH protocol between the users who are under the left child node and the users who are under the right child node from bottom to up.

For contributory key management schemes, the number of rounds is usually used to measure the communication, computation, and latency [120] associated with key establishment and updating [58,66,67].

With the same simulation setup as that in Section 7.4.2, the performance of the independent-tree and multi-group contributory key management schemes are compared for varying group size. Figure 7.14 shows the total number of rounds to establish the group key, which reflects the latency in key establishment [120]. Figure 7.15 shows the number of rounds performed by the users in each SG, which describes the users' computation overhead. In each round, a user performs two modular exponentiations. With the same simulation setup as that in Section 7.4.2, Figure 7.16 shows the number of rounds for key updating for with different number of

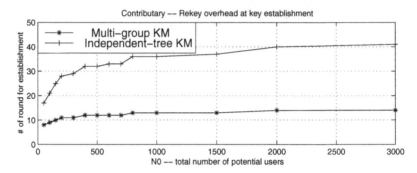

FIGURE 7.14. The total number of rounds performed to establish the group key

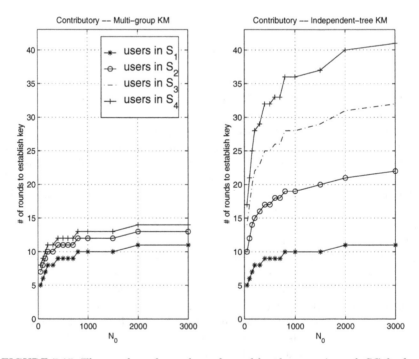

FIGURE 7.15. The number of rounds performed by the users in each SG for key establishment

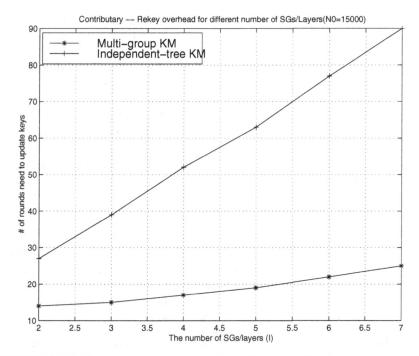

FIGURE 7.16. The number of rounds performed to establish the group key with different number of SGs/layers

layers. Compared with the tree-based contributory schemes, the multi-group contributory scheme significantly reduces the computation and latency associated with key establishment and updating. The advantage of the multi-group contributory scheme is larger when M increases.

7.6   Related Work

In this chapter, we have presented key management for Hierarchical Group Access Control. In the research literature, there is a similar but different problem called key management for access hierarchy. For make the presentation concise, we use HGAC to represent Hierarchical Group Access Control and KMAH to represent Key Management for Access Hierarchy. Next, we briefly introduce KMAH, and describe the difference and relationship between KMAH and HGAC.

In KMAC, there are a set of access classes that forms an access hierarchy. The access hierarchy is usually represented by a directed graph where nodes correspond to access classes and edges indicate their ordering. If a user is entitled to have access to a certain access class, he/she is also entitled to access the descendant classes. The key management scheme for the access hierarchy assigns keys to the access classes and distributes a subset of the keys to users, which permits the users to access the files belonging to a particular access class and all of the descendant classes.

In this type of key management schemes, the keys of the descendant classes are often derived from the keys of the higher classes in the hierarchy. The main performance measure is the number of operations needed to compute the keys of the descendant classes. The KMAC is often applied to scenarios where deep access hierarchies are needed, such as in a database where very complicated access rules are posed. Currently, all approaches assume existence of a central authority. A nice summary of existing works on access hierarchy can be found in [121].

We summarize the differences between KMAC and HGAC as follows.

- In KMAH, the access rules can be very complicated even for a simple database. As a result, the design of KMAH focuses on managing keys associated with the graph that connects the access classes. This graph represents the relationship between access classes. The major challenge is to design an efficient way to derive keys from the descendant classes. KMAH can be used for group communications, but not mainly designed for group communications.

- In HGAC, the access relationship is represented by a set of DG-subtrees. In group communications, the access relationship is usually not complicated. Thus, simple structures can be used for each DG-subtree, such as a binary tree. All DG-subtrees together can represent the access relationship. On a DG-subtree, the nodes are the KEKs, and edges describe the ownership of the KEKs. When the group size is large, the overhead is mainly introduced by updating the keys on the SG-subtree. Therefore, HGAC must consider the design of both the DG-subtrees and the SG-subtrees. The major challenge is to design good key tree structures and key updating protocols for the purpose of reducing overhead when the group membership is highly dynamic.

From the above discussion, we can also see the relationship between KMAH and HGAC. That is, some techniques in KMAH can be used to design the DG-subtrees in HGAC. In [122], the partial order of service groups is considered into the design of the DG-subtrees. As a consequence, the storage overhead and rekey overhead are reduced, especially when the group size is small. Exploiting more techniques in KMAH could result in further reduction in the overhead of the HGAC solutions.

7.7 Chapter Summary

The focus of this chapter is formulating and solving the hierarchical access control problem in group communications. The goal is to assure forward and backward secrecy when multiple data streams are distributed to group members with varying access privileges. Meanwhile, the overhead should be kept low. Two solutions, the multi-group scheme and the independent-tree scheme, are described and compared. The former outperform the later in terms of communication, computation and storage overhead. The extensions of the two solutions in the contributory environment are also discussed.

8

Protecting Membership Information in Secure Multicasting

Many existing key management schemes focus on maintaining key secrecy and reducing the communication overhead associated with updating the associated keys [78] [7] [56]. However, it is found that key management can disclose information about dynamic group membership to both insiders and outsiders. In other words, while the content of group communication is protected by encryption using the secret keys, group dynamic information is disclosed through key management. *Group dynamic information* (GDI) is the information that describes the dynamic group membership, including the number of users in a multicast group as a function of time, and the number of joining or departing users in a time interval.

In many secure group applications, group dynamic information should be kept confidential [123, 124]. Key management is a technology that enables key updating in real time as group membership changes. Future commercial multicast services, which could occur in non-traditional broadcast media such as Internet and 3G/4G wireless networks, will allow a user to subscribe to an arbitrary set of programs and change his/her subscription at any time [125] [10]. The users can choose to pay for exactly what they get, instead of a fixed monthly fee. This new type of services give the most flexibility to users, as well as opportunities to new business models. Over the non-traditional broadcast media, the global media giants as well as small multimedia producers can be the service providers. The service providers perform group management and have the knowledge of GDI, i.e audience statistics. However, it is highly undesirable to disclose instant and detailed GDI to competitors. Assume a competitor can monitor the audience statistics of the service provider X. Then, the competitor may broadcast its programs at different time slots and see how it affects its own and X's audience statistics. As a consequence, the competitor can develop the best program schedule to compete with X. This example also shows that *GDI should also be concealed from insiders*. A regular user, who receives the multicast content, should not know the overall audience statistics. Otherwise, the competitor can send one of its employees to register as X's member for a small cost, and collect valuable audience statistics from X. In addition, there are multicast communication scenarios where GDI represents sensitive deployment information about the network. For example, in a sensor network, the base station sends many broadcast messages to sensors. The base station and sensors form a secure multicast group. If some sensors are compromised, the group key should be updated such that the compromised sensors cannot decrypt future multicast messages from the BS. One possible way to update group keys is to use group key management schemes. In such an application scenario, GDI represents the number of sensors deployed in an area, and the number of revoked sensors. In this example, if GDI is not protected, attackers can obtain sensor deployment information by exploiting the key management scheme.

From the above two examples, one can see that revealing GDI through key management could be a new type of vulnerability. In this chapter, the focus is to investigate GDI leakage problem [123, 124] and to present a framework of protecting GDI from insiders and outsiders.

8.1   GDI Disclosure in Centralized Key Management Schemes

In the centralized key management schemes, there exists a KDC that generates and distributes the decryption keys [78]. In this section, we investigate the methods that can acquire GDI stealthily from the centralized key management.

The group dynamic information (GDI) particularly refers to a set of functions as:

- $N(t)$: the number of users in the multicast group at time t.

- $J(t_0, t_1)$: the number of users who join the service between time t_0 and t_1.

- $L(t_0, t_1)$: the number of users who leave the service between time t_0 and t_1.

The GDI should be kept confidential in many group-oriented applications, yet to acquire GDI from key management can be simple and stealthy. Instead of trying to break encryption or compromise the key distribution center, the adversaries can subscribe to the service as regular users. In

this case, they are referred to as *insiders*. Insiders can obtain very accurate estimation of GDI by monitoring the *rekeying messages*, which are the
messages conveying new key updating information. Even if the adversaries
cannot become valid group members, they can still obtain GDI as the *outsiders* as long as they can observe the rekeying traffic around a single group
member.

This section presents three methods to obtain GDI based on the key
management scheme proposed in [80], and discuss the vulnerability of other
prevalent centralized key management schemes.

Although having different rekeying procedures, most tree-based centralized key management schemes [7, 8, 10, 78–80] share several common properties. First, group members can distinguish the key updating process due
to user join and that due to user departure. Second, the rekeying-message-
size may be related with the group size. Third, the IDs of the keys stay
the same even if the key content changes. Because of these properties, several methods can be used to obtain GDI stealthily from key management.
Next, those methods are presented based on the tree-based key management
scheme in [80].

8.1.1 Attack 1: Estimation of $J(t_0, t_1)$ and $L(t_0, t_1)$ from rekeying-message format

To demonstrate the attacks, we consider the popular tree-based centralized
key management scheme proposed in [80]. In brief, when a user leaves the
group, all the keys on the path from this user to the root of the key tree
are updated by conveying a set of rekeying messages, that have the basic
format as one key encrypted by another key (see Section 5.1). When a user
joins the service, the KDC chooses a leaf position on the key tree to put the
joining user. The KDC updates the keys along the path from the new leaf
to the root by generating the new keys from the old keys using a one-way
function and increasing the revision numbers of the new keys. (Each key
is associated with a revision number indiating the version of the key [80].)
The joining user obtains the new keys through the unicast channel. Other
users in the group will know about the key change when the data packet
indicating the increase of the revision numbers first arrives, and compute
the new keys using the one-way function. No additional rekeying messages
are necessary.

An insider receives rekeying messages, decrypts some of the messages,
and observes the rekeying-message-size without having to understand the
content of all messages. Since the key updating process for user join and
the process for user departure are different, he can estimate $J(t_0, t_1)$ and
$L(t_0, t_1)$ as follows.

- When receiving the rekeying message containing K_s^{new} encrypted by
  one of his KEKs, he assumes that one user leaves the group.

- When observing the increase of the revision number of K_s, he assumes that one user joins the group.

(The detailed rekeying message format can be found in Section 5.1.)

This strategy is effective when most users do not join/leave simultaneously and the keys are updated immediately after each user joining/departing event. When this method is successful, $N(t)$ can be calculated from $J(t_0, t_1)$ and $L(t_0, t_1)$ as:

$$N(t_1) = N(t_0) + J(t_0, t_1) - L(t_0, t_1). \tag{8.1}$$

Even if the initial group size is unknown, the changing trend of the group size is obtained.

8.1.2 Attack 2 : Estimation of the group size from the rekeying-message-size

In some tree-based key management schemes [126], key tree is fully loaded and maintained as balanced as possible by putting the joining users on the shortest branches. In this case, the group size $N(t)$ can be estimated directly from the rekeying-message-size.

It is assumed that $N(t)$ does not change much within a short period of time. In this time period, there are W departing users who do not leave simultaneously. Thus, W observations of the rekeying-message-size due to single user departure are made. These observations are denoted by $Msg = \{m_1, m_2, \cdots, m_w\}$.

In the worst-case scenario, the insiders and outsiders know the degree of the key tree, denoted by d. Then, they can calculate the length of the branch where the ith leaving user was located before his departure, denoted by L_i. Without losing information, the observed Msg is converted to $\{L_1 = l_1, L_2 = l_2, \cdots, L_W = l_W\}$, where $l_i = \lceil \frac{m_i + 1}{d} \rceil$. Then, the Maximum Likelihood (ML) estimator of the group size is formulated as:

$$N_{ML} = \arg\max_n \ Prob\{L_1 = l_1, L_2 = l_2, \cdots, L_W = l_W | N(t) = n\}, \tag{8.2}$$

where $\arg\max_n g(n)$ represents the n value that maximizes the function $g(n)$. To solve (8.2), it is necessary to introduce a set of new variables: $\{S_k\}_{k=L_{min}, L_{min}+1, \cdots, L_{max}}$, where S_k is the number of users who are on the branches with length k, L_{max} is the length of the longest branches, and L_{min} is the length of the shortest branches. It is obvious that

$$\sum_k S_k = n. \tag{8.3}$$

In addition, the length of the branches of a key tree must satisfy the Kraft inequality [30], i.e. $\sum_j d^{L_{max} - b_j} \leq d^{L_{max}}$, where b_j is the length of the branch on which the user j stays and $j = 1, 2, \cdots, n$. Thus, S_k, which equals to the number of elements in set $\{b_j : b_j = k\}$, must satisfy

$$\sum_k S_k d^{L_{max} - k} \leq d^{L_{max}}, \tag{8.4}$$

It can be verified that the equality is achieved when all intermediate nodes on the key tree have d children nodes. When the key tree is balanced and fully loaded, it is reasonable to approximate (8.4) by

$$\sum_k S_k d^{L_{max} - k} = d^{L_{max}}. \tag{8.5}$$

When the leaving users are uniformly distributed on the key tree, and the number of users in the system is much larger than the number of leaving users, i.e. $N(t) >> W$. Then, the probability mass function (pmf) of L_i is

$$Prob\{L_i = k \mid n, \ S_k\} = \frac{S_k}{n}, \quad k = L_{min}, L_{min} + 1, \cdots, L_{max}.$$

It is assumed that $L_i, i = 1, \cdots, W$ are i.i.d. random variables. Thus, the probability in (8.2) is calculated as:

$$Prob\{L_1 = l_1, L_2 = l_2, \cdots, L_W = l_W \mid N(t) = n, \ S_k\} = \prod_k \left(\frac{S_k}{n}\right)^{h(k)}, \tag{8.6}$$

where $h(k)$ denotes the number of elements in set $\{l_i : l_i = k\}$ and obviously, $\sum_k h(k) = W$. Then, the values of n and $\{S_k\}$ that maximize (8.6) under the constraint (8.3) and (8.5) are obtained using Lagrange multiplier as:

$$\{S_k\}_{ML} = \frac{n}{W} h(k), \tag{8.7}$$

$$N_{ML} = \frac{W}{\sum_k h(k) d^{-k}}. \tag{8.8}$$

Next, the performance of this ML estimator is demonstrated through a set of simulations. The estimator is first applied to *simulated* dynamic group membership changes. As suggested in [90] [72], the user arrival process is modeled as a Poisson process, and the service duration is modeled as an exponential random variable. In Figure 8.1(a), 8.1(b), and 8.1(c), the estimated group size is obtained by using the estimator in (8.8), and compared with the true values of $N(t)$. These plots are for three different simulation settings. The entire service period is divided into four sessions. The model parameters, i.e. the user arrival rate and the average service time, are fixed within each session and vary in different sessions. In the ith session, described by interval $[t_{i-1}, t_i)$, the user arrival rate is λ_i and the average service time is μ_i. In all three cases, $[t_0, t_1, t_2, t_3, t_4]$ is chosen

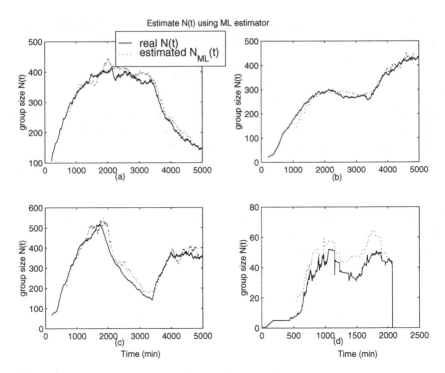

FIGURE 8.1. Performance of the ML estimator. (a)(b) and (c) are for simulated multicast sessions. (d) is for a Mbone session.

to be $[0, 200, 1600, 3200, 5000]$ minutes, and the initial group size is 0. The parameter λ_i's and μ_i's as follows.

	$[\lambda_1, \lambda_2, \lambda_3, \lambda_4] =$	$[\mu_1, \mu_2, \mu_3, \mu_4] =$
plot(a)	$[0.5, 0.5, 0.5, 0.3]\mathrm{min}^{-1}$	$[1400, 800, 600, 400]\mathrm{min}$
plot(b)	$[0.1, 0.3, 0.2, 0.5]\mathrm{min}^{-1}$	$[1500, 1500, 1000, 800]\mathrm{min}$
plot(c)	$[0.3, 0.7, 0.1, 0.9]\mathrm{min}^{-1}$	$[1400, 800, 600, 400]\mathrm{min}$

Figure 8.1(d) demonstrates the performance of the ML estimator, when it was applied to a real *MBone* audio session, CBC Newsworld on-line test, started on Oct. 29. 1996 and lasted for about 5 days [127].

In all four cases, the changing trend of the group size is well captured by the estimator. It is also observed that the estimated group size tends to be larger than the true $N(t)$. This is due to the approximation when replacing (8.5) by (8.4). Although not perfect, this estimator is effective for analyzing audience behavior and the group size changes.

8.1.3   Attack 3: Estimation of group size based on key IDs

As presented in [80], each key contains the secret material that is the content of the key and a *key selector* that is used to distinguish the key. The key

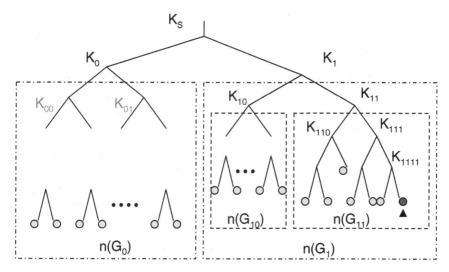

FIGURE 8.2. Key ID based attack method

selector consists of: 1) a unique ID that stays the same even if the key content changes and 2) a version and revision field, reflecting update of the key. The basic format of the rekeying messages is $\{K_y\}_{K_x}$, representing K_y encrypted by K_x. This message has two parts. The first part is the key selector of K_x, which is not encrypted because otherwise a user will not be able to understand this message. The second part is K_y and the key selector of K_y, encrypted by K_x. Thus, in the current implementation, everyone who can overhear the rekeying messages can see the IDs of K_x.

One can collect the histogram of these key IDs. Let $P(K_x)$ denote the probability of K_x's ID appears, calculated as the number of rekeying messages containing this ID (as an encryption key ID) divided by the total amount of rekeying messages. Define G_x as the set of users under the node associated with K_x, and $n(G_x)$ as the number of users in G_x. Let K_p denote the parent node of K_x.

Based on the rekey procedure in [80], one can see that $P(K_x)$ equals to the probability that one or more than one users leave the subgroup G_p given that there are users leaving the multicast group. In addition, it is reasonable to assume that $P(K_x)$ is proportional to $n(G_p)$. This assumption is valid when users are equally likely to leave and the probability of a user leaving in one round of key updating is small.

Above discussion leads to the third attack: *Key ID based attack*. The basic idea of this attack is explained using the example shown in Figure 8.2, where the attacker is marked by a triangle.

- Step 1: The attacker knows that the keys on the branch from himself to the root are $\{K_s, K_1, K_{11}, K_{111}, K_{1111}\}$. Among these keys, he also knows who is whose children node because the parent node keys are always encrypted by the children node keys. The attacker collects $P(K_1)$, $P(K_{11})$, and $P(K_{111})$ by observing a sufficient number of rekeying messages.

- Step 2: When there are users leaving G_0, KDC needs to update key K_0 by sending rekeying message $\{K_0\}_{K_{00}}$ and $\{K_0\}_{K_{01}}$, according to the rekeying procedure described in Section II-A. Thus, whenever there are users leaving G_0, the IDs of K_{01} and K_{00} will appear. Therefore, $P(K_{00}) = P(K_{01}) = Pr(\text{there are users leaving } G_0)$. Similarly, we have $Pr(\text{there are users leaving } G_1) = P(K_{10}) = P(K_{11})$ and $Pr(\text{there are users leaving } G_s) = P(K_0) = P(K_1)$. Since $G_s = G_0 \bigcup G1$ and $G_0 \bigcap G_1 = \phi$, it is easy to see that $P(K_1)$ is just $Pr(\text{there are users leaving } G_1 \text{ or there are users leaving } G_0)$. Therefore, $P(K_1)$ is calculated as

$$P(K_1) = 1 - (1 - P(K_{11}))(1 - P(K_{00})). \tag{8.9}$$

Here, since $P(K_{11}) = P(K_{10})$ and $P(K_{00}) = P(K_{01})$, one can replace $P(K_{11})$ by $P(K_{10})$ and/or replace $P(K_{00})$ by $P(K_{01})$ in (8.9).

In addition, as described earlier, it is reasonable to assume that

$$\frac{n(G_0)}{n(G_1)} = \frac{Pr(\text{there are users leaving } G_0)}{Pr(\text{there are users leaving } G_1)} = \frac{P(K_{00})}{P(K_{11})}. \tag{8.10}$$

Similarly, the attacker can obtain $\frac{n(G_{10})}{n(G_{11})}$, and $\frac{n(G_{110})}{n(G_{111})}$.

- Step 3: The attacker estimates $n(G_{111})$ based on the degree of the key tree. Then, he can obtain $n(G_{110})$, $n(G_{10})$, and $n(G_0)$, using the results generated in the previous step. The group size is finally estimated as $n(G_{111}) + n(G_{110}) + n(G_{10}) + n(G_0)$.

The accuracy of this attack depends on the estimation error of $n(K_{111})$. In this example, $n(K_{111})$ can be estimated as either 3 or 4. This results in 25% estimation error in the total group size. The accuracy also depends on the assumption that group members are equally likely to leave and they leave independently. Although it is not a very accurate method, key ID based attack can reveal a large amount of GDI information.

More importantly, equation (8.9) and (8.10) do not rely on specific tree structures. When the key tree is not balanced and/or not fully-loaded, those equations are still valid. To see this, one can examine an example of an unbalanced key tree, where there are N users under K_1 and $N/5$ users under K_0. When the departuring users are randomly located on the key tree, $Pr(\text{there are users leaving } G_1) \approx 5 \cdot Pr(\text{there are users leaving } G_0)$. Therefore, the ID of key K_{11} should appears 4 times more frequently than the ID of key K_{00}. Using the procedure in step 2, the attacker can know that the number of users under K_1 are approximately 4 times more than the users under K_0 by examining the key IDs. Thus, Attack 3 can be applied to unbalanced or non-fully loaded key trees. This is the major advantage of Attack 3. Recall that Attack 2 is suitable for balanced and fully-loaded key trees.

8.1.4 Discussion on three attacks

An insider can jointly use all three types of attacks, and an outsider can use AII and AIII under certain conditions. An outsider can apply AII when he is able to observe the size of rekeying messages. It has been shown that the rekeying messages must be delivered reliably and in a timely manner, in order to guarantee the quality of service [128]. Therefore, it is very likely that rekeying messages are treated differently from the regular data in terms of error control, or even transmitted in a reliable multicast channel that is separated from the channel used for data transmission. This provides an opportunity for the outsiders to differentiate the rekeying messages and the multicast content. As long as an outsider can observe the rekeying traffic sent to one group member, he can obtain the rekeying-message-size and use method AII to estimate the group size. It is noted that error control coding may change the size of the rekeying messages. The coding rate is often not a secret. Thus, the attackers can recover the original rekeying message size before coding. In current key management schemes, the key selector of the encryption key is not encrypted. Thus, an outsider can collect the histogram of key IDs. One straightforward improvement is to use the session key to encrypt the key selector, which will prevent outsiders from using AIII. This requires additional encryption/decryption operations.

In the derivation of the ML estimator in Attack 2, it is assumed that the key tree is fully loaded. This assumption can be violated in some implementations of key management. For example, the KDC first estimates the maximum group size to be N_{max}. Then, a key tree with N_{max} leaf nodes is constructed. This key tree will have many empty leaf nodes that are not associated with particular users. A joining user will occupy an empty leaf node after it joins, and a departing user will release a leaf node after it leaves. Since there is no need to split or merge nodes when users join or leave, these type of key trees are easy to maintain. On the other hand, they often require higher overhead to store and update keys than what is necessary. In practice, the type of key trees, referred to as non-fully loaded key trees, are used when N_{max} is not large or the difference between N_{max} and the average group size is not large. For non-fully loaded key trees, Attack 3 should be applied. Although the accuracy of Attack 3 is not as good as other attacks, it still can provide a large amount of information about GDI. If multiple attackers jointly estimate GDI, the results can be more accurate.

As a summary, the properties of three attacks are listed in Table 8.1.

TABLE 8.1. Comparison among attack methods. (* when the initial group size is known.)

	applied by insider	applied by outsider	requirement on key trees	accuracy
A1	yes	no	none	high*
A2	yes	possible	balanced, full-loaded	high-moderate
A3	yes	possible	none	moderate-low

8.1.5   GDI vulnerability in prevalent key management schemes

While Attack 3 is only suitable for tree-based schemes, Attack 1 and 2 can be tailored to many other key management schemes. When the insiders can differentiate the rekeying messages for user join and those for user departure, they use an attack similar to AI, referred to as the *AI type method*. When the amount of rekeying messages largely depends on the group size, they can use an attack similar to AII, referred to as the *AII type method*, with an estimator that may be different from (8.8). In this section, the vulnerabilities of popular centralized and decentralized key management schemes are reviewed.

Since protecting GDI is not a part of the design goal of traditional key management schemes, it is not surprising that some schemes reveal GDI in a very straightforward way. For example, in the approach proposed in [129], a security lock is implemented based on the Chinese remainder theorem and the length of the lock is proportional to the number of users. Thus, $N(t)$ is obtained by measure the length of the lock, which is the simplest AII type method.

Tree-based key management schemes have been known for their efficiency in terms of communication, computation and storage overhead. Many tree-based schemes, such as [7, 8, 79], are similar to that described in Section 8.1.1. In these cases, both the AI and AII methods can be applied. In [10, 117, 118], another class of tree-based schemes were presented to further reduce the communication overhead by introducing the dependency among keys, such as in one-way function trees. In these schemes, the key updating procedures for user join and departure are similar. Thus, AI type methods are not applicable. Since the size of rekeying messages is closely related with the group size, AII type methods are still suitable.

Besides the tree-based scheme described in Section 5.1, the VersaKey framework [80] also includes a centralized flat scheme. When a user joins or leaves the group, the rekeying-message-size equals to the length of the binary representation of user IDs, which can be independent of $N(t)$. Thus, this key management scheme is resistant to both the AI and AII type methods. This scheme, however, is vulnerable to collusion attacks. That is, the KDC cannot update keys without leaking new key information to the leaving user, who has a collusion partner in the group. Although the

GDI is protected, this scheme does not protect the multicast content when collusion attacks are likely.

In Iolus [130], a large group is decomposed into a number of subgroups, and the trusted local security agents perform admission control and key updating for the subgroups. This architecture reduces the number of users affected by key updating resulting from membership changes. Since the key updating is localized within each subgroup, the insiders or outsiders can only obtain the dynamic membership information of the subgroups that they belong to or can monitor.

The idea of clustering was introduced in [131] to achieve the efficiency by localizing key updating. The group members are organized into a hierarchical clustering structure. The cluster leaders are selected from group members and perform partial key management. Since the cluster leaders establish keys for the cluster members through pair-wise key exchange [131], the cluster members cannot obtain GDI of their clusters. However, the cluster leaders naturally obtain the dynamic membership information of their clusters and all clusters below. In [131], the cluster size is chosen from 3 to 15. Therefore, this key management scheme can be applied only when a large portion of group members are trusted to perform key management and obtain GDI.

In Chapter 5, a topology-aware key management (TMKM) scheme was presented. It reduces the communication overhead by matching the key tree with the network topology and localizing the transmission of the rekeying messages. In this scheme, group members receive only the rekeying messages that are useful for themselves and their neighbors. Thus, they only obtain the local GDI by using AI or AII type methods.

As a summary, Table 8.2 lists various key management schemes discussed in this section. We can see that the AII type methods are effective for obtaining GDI or local GDI from many key management schemes. Two schemes, flat VersaKey [80] and the clustering in [131] do not reveal GDI, but their usage are limited because they are either not resistant to collusion attacks or must put trust in a large number of cluster leaders. Therefore, the defense techniques that protect GDI should be compatible with a variety of key management schemes.

8.2 Defense Techniques

The discussion on GDI attacks in this chapter does not cover all aspects of key management schemes that can reveal group dynamic information. New attacks may emerge in the future. Therefore, the defense mechanism should be robust against various threats and compatible with different key management schemes.

The rekeying process reveals GDI in two domains. In the time domain, the insiders/outsiders observe when the rekeying messages are transmitted.

TABLE 8.2. Vulnerability of prevalent key management schemes

Centralized Key Management Schemes		Is AII Effective?	Is AI Effective?
Tree based	Key Graph [8] Wallner98 [7] VersaKey [80] Embedding [79]	Yes	Yes
	One-way function tree [117] Improve key revocation [10] ELK [118]	Yes	No
Flat	Security lock [129]	Yes	–
	Flat centralized scheme VersaKey [80]*	No	No
Local security agents	Iolus [130]	Local	Local
	Clustering [131]*	No	No
Others	TMKM [132]	Local	Local

In the message domain, the insiders/outsiders observe the size and/or the format of the rekeying messages.

To protect GDI in the time domain, *batch rekeying* [80] [119] is an effective method. Batch rekeying postpones the updates of the keys, and can remove the correlation between the time of key updating and the time when users join/leave the group. In particular, batch rekeying is often implemented as periodic updates of keys. The users who join or leave the group in the time interval $[(k-1)B_t, kB_t]$, are added to or removed from the key tree together at time kB_t, where k is an positive integer and B_t is the key updating period. By doing so, the time-domain observations do not contain information about when users join/leave the group. It is important to note that batch rekeying was originally proposed to reduce the rekeying overhead. It has been shown in [80, 119, 133] that updating keys for several users together consumes less communication and computation resources than updating keys for the users one-by-one. The disadvantage of batch rekeying is that the joining/departing users will be able to access a small amount of information before/after their join/departure. Thus, the parameter B_t must be chosen based on the group policies. In particular, B_t should be smaller than the maximum acceptable delay between revoking a user and sending information that should not be accessed by the revoked user. When using batch rekeying, the notations of the GDI functions are simplified as: $J(k) = J((k-1)B_t, kB_t)$, $L(k) = L((k-1)B_t, kB_t)$, and $N(k) = N(kB_t)$.

Batch rekeying cannot protect GDI in the message domain. Figure 8.3 shows simulation results for the batch rekeying when B_t is set to be 5 minutes. Simulation setup is similar to that in Section 8.1.2. The solid line in Figure 8.3(a), 8.3(b), 8.3(c), 8.3(d) represent the $N(k)$, $J(k)$, $L(k)$

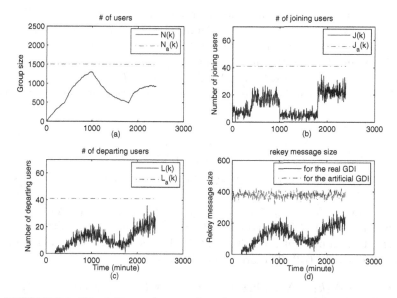

FIGURE 8.3. The defense scheme using phantom users and Batch rekeying

and the rekeying-message-size, respectively. One can see that the rekeying-message-size is closely related to $L(k)$ and reflects the trend of $N(k)$. A large amount of information about $N(k)$ and $L(k)$ is in the message domain.

To reduce the amount of GDI in the message domain, an effective method is to insert phantom users into the system. These phantom users, as well as their join and departure behaviors, are created by the KDC in such a way that the combined effects of the phantom users and the real users lead to a new rekeying process, called the *observed rekeying process*.

Let $N_a(k)$ denote the total number of the real and phantom users, and $J_a(k)$ and $L_a(k)$ denote the total number of the real and phantom users who join/leave the group respectively. $N_a(t)$, $J_a(k)$, and $L_a(k)$ are referred to as the *artificial GDI*. From the key management points of view, the phantom users are treated just as the real users. They occupy leaf nodes on the key tree, and they are associated with a set of KEKs that are updated when they virtually join or leave the group. Thus, the observed rekeying process only depends on the artificial GDI.

There are many choices for artificial GDI functions. The simplest candidates are constant functions.

$$J_a(k) = L_0, \quad L_a(k) = L_0, \quad N_a(k) = N_0. \tag{8.11}$$

If the constant artificial GDI functions could be implemented, the observed rekeying process would not leak the information about the changing trend of the real GDI. However, the perfect flat artificial GDI functions in (8.11) may not be achievable. Since the real GDI functions are random processes, it is possible that the predetermined L_0 and N_0 are not large enough such that the artificial GDI cannot be maintained as the straight lines. For example, when $N(k) > N_0$, $N_a(k)$ cannot be N_0 because the number of phantom users must be non-negative. In fact, the artificial GDI functions must satisfies four requirements: (r1) $N_a(k) \geq N(k)$, (r2) $L_a(k) \geq L(k)$, (r3) $J_a(k) \geq J(k)$, and (r4) $N_a(k) = N_a(k-1) + J_a(k) - L_a(k)$.

A modification can be made to the constant functions such that (r1)-(r4) are satisfied.

$$N_a(k) \;=\; \max\{N(k), N_0\} \tag{8.12}$$
$$J_a(k) \;=\; \max\{J(k), L(k), L_0\} \tag{8.13}$$
$$L_a(k) \;=\; N_a(k-1) - N_a(k) + J_a(k) \tag{8.14}$$

When $N(k) \leq N_0$, $L(k) \leq L_0$, and $J(k) \leq L_0$, equation (8.12)-(8.14) are equivalent to (8.11). The artificial GDI functions in (8.12)-(8.14) obviously satisfy the requirement (r1) (r3) and (r4). The modified artificial GDI functions also satisfy (r2). The proof is as follows.

- When $N(k) > N_0$, using the fact that $N_a(k-1) \geq N(k-1)$, $N_a(k) = N(k)$, and $J_a(k) \geq J(k)$, one can see that $L_a(k) = N_a(k-1) - N_a(k) + J_a(k) \;\geq\; L(k) = N(k-1) - N(k) + J(k)$.

- When $N(k) \leq N_0$, using the fact that $N_a(k-1) \geq N_0$ and $J_a(k) \geq L(k)$, we get $L_a(k) \geq J_a(k) \geq L(k)$.

It should be noted that there are many other ways to choose the artificial GDI functions. Some artificial GDI functions can protect GDI better than others. Artificial GDI functions can also be non-deterministic. In this chapter, we use the artificial GDI functions in (8.12)-(8.14) to demonstrate the defense mechanism. Searching for the better artificial GDI functions can be an interesting future research problem.

Given the artificial GDI functions, the KDC creates phantom users and performs key management as follows.

(1) Determine N_0 and L_0 based on the system requirements and the users' statistical behavior. The criteria for selecting N_0 and L_0 will be presented in Section 8.3.

(2) Before the group communication starts, create N_0 phantom users and establish a key tree to accommodate them. Set index $k = 1$.

(3) While the communication is not terminated, execute the follows.

   – Record user join and departure requests in the time period $((k-1)B_t, kB_t]$, and obtain $J(k)$ and $L(k)$. During this time, the current session key is sent to the joining users such that they can start receiving the multicast content without delay.

- At time kB_t, the KDC creates $J_a(k) - J(k)$ phantom users joining the service, and then selects $L_a(k) - L(k)$ phantom users
  in the current system and makes them leave. Following the key
  updating procedure presented in any existing key management
  schemes, the KDC updates corresponding keys for real and phantom users' join and departure. The number of total real and
  phantom users are maintained to be $N_a(k)$.

- Set $k = k + 1$.

Figure 8.3(a), 8.3(b), and 8.3(c) illustrate the real GDI $(N(k)$, $L(k)$,
$J(k))$ and the artificial GDI $(N_a(k)$, $L_a(k)$, $J_a(k))$ for a simulated multicast service. The simulation results of the communication overhead, i.e. the
rekeying-message-size, is shown in Figure 8.3(d). Here, the solid line represents the case with batch rekeying but no phantom users. The dashed line
represents the case when the proposed defense method is applied. It is important to note that batch rekeying technique is used for all results shown
in Figure 8.3. One can see that the observed rekeying process reveals very
limited information about the real GDI when the proposed defense scheme
is used. The rekeying message size resulting from using batch rekeying along
is still highly correlated with the group size. In addition, the communication
overhead increases, which is a disadvantage of utilizing phantom users.

Utilizing phantom users and batch rekeying is not the only solution to the
problem of GDI leakage. There are other techniques that can protect GDI
against one or several attack methods. For example, to prevent outsiders
from launch the AII type attack, the rekeying messages can be embedding
into multicast content [79] or transmitted using onion-routing [134]. Using
the same rekeying procedure for user join and departure is also a good
way to prevent the AI type attacks. In addition, the KDC can generate
fake rekeying messages to prevent the AII type methods. The fake rekeying
message could have a header indicating it is a rekeying message but the
content is random bits. This is different from the proposed defense scheme
where the key tree reserves slots for the phantom users and all rekeying
messages have meanings. Compared with other techniques, using phantom
users and batch rekeying has two major advantages. First, the proposed
defense scheme is effective against various attacks. Since the real GDI is
concealed *before* the rekeying messages are generated and even before key
selectors are modified, only the artificial GDI can be seen from the observed
rekeying process unless the KDC is compromised. Second, the proposed
scheme does not rely on specific rekeying algorithms and is compatible
with existing key management schemes.

8.3 Optimization

The idea of employing phantom users is not complicated. The challenge is to determine the amount of phantom users that should be inserted. In this section, an optimization problem is formulated such that the proper amount of phantom users, described by the parameter L_0 and N_0 in (8.12)-(8.14), can be determined.

Before discussing the optimization problem, the performance criteria must be determined. In this section, two criteria are defined. They are (a) the amount of information leaked to the insiders and outsiders measured by mutual information, and (b) the communication overhead introduced by the phantom users. The tradeoff between these two metrics will be studied.

8.3.1 The leakage of GDI

For measuring the leakage of the GDI, a nature measure is mutual information. The mutual information between two random variables X and Y describes how much information that one can tell about X when knowing Y.

Let T be the total number of rounds of key updates. The overall service duration is $T \cdot B_t$. Then, the real GDI is described by a set of random variables as

$$R \;=\; \{N(1), \cdots, N(T), J(1), \cdots, J(T), L(1), \cdots, L(T)\}, \quad (8.15)$$

and the artificial GDI is

$$A \;=\; \{N_a(1), \cdots, N_a(T), J_a(1), \cdots, J_a(T), L_a(1), \cdots, L_a(T)\}. \quad (8.16)$$

The mutual information, $I(R; A)$, describes the reduction in the uncertainty of the real GDI due to the knowledge of the artificial GDI [30]. Therefore, the leakage of the GDI can be measured by

$$I(R; A) = H(A) - H(A|R), \quad (8.17)$$

where $H(.)$ and $H(.|.)$ denote the entropy and conditional entropy, respectively.

Equation (8.12) - (8.14) indicate that the artificial GDI is a set of deterministic functions of the real GDI. Thus, the conditional entropy in (8.17) is zero, i.e. $H(A|R) = 0$. Since $L_a(k)$ is directly computed from $J_a(k)$, $N_a(k)$ and $N_a(k-1)$ in (8.14), the terms $L_a(1), L_a(2), \cdots, L_a(T)$ can be removed from the expression of the entropy of A, i.e. $H(A) = H(N_a(1), \cdots, N_a(T), J_a(1), \cdots, J_a(T))$. Then, the upper bound of $I(R; A)$ is calculated as:

$$
\begin{aligned}
I(R; A) \;&=\; H(N_a(1), \cdots, N_a(T), J_a(1), \cdots, J_a(T)) \\
&\leq\; \sum_k H(N_a(k)) + \sum_k H(J_a(k)). \quad (8.18)
\end{aligned}
$$

The equality is achieved when $\{N_a(k), J_a(k), k = 1, \cdots, T\}$ are mutually independent. It is noted that the GDI at time kB_t and the GDI at time $(k+1)B_t$ can be approximately independent when B_t is large and the group is highly dynamic. In these cases, (8.18) provides a tight upper bound of $I(R; A)$.

Let $p_{N_k}(n)$ and $p_{N_{ak}}(n)$ denote the probability mass function (pmf) of $N(k)$ and $N_a(k)$, respectively. From (8.12), one can see that

$$
p_{N_{ak}}(n) = \begin{cases} \sum_{x=0}^{N_0} p_{N_k}(x), & n = N_0 \\ p_{N_k}(n), & n > N_0 \\ 0, & o.w. \end{cases}
$$

Then,

$$
H(N_a(k)) = -(1 - \epsilon_N^k) \log(1 - \epsilon_N^k) - \sum_{n=N_0+1}^{\infty} p_{N_k}(n) \log p_{N_k}(n), \quad (8.19)
$$

where $\epsilon_N^k = 1 - \sum_{x=0}^{N_0} p_{N_k}(x)$. Similarly, let $p_{J_k}(x)$, $p_{J_{ak}}(j)$, and $p_{L_k}(y)$ denote the pmf of $J(k)$, $J_a(k)$, and $L(k)$, respectively. Thus,

$$
H(J_a(k)) = -\sum_j p_{J_{ak}}(j) \log p_{J_{ak}}(j), \quad (8.20)
$$

and,

$$
p_{J_{ak}}(j) = \begin{cases} (1 - \epsilon_J^k)(1 - \epsilon_L^k), & j = L_0 \\ p_{J_k}(j) \sum_{y=0}^{j-1} p_{L_k}(y) \\ \quad + p_{L_k}(j) \sum_{x=0}^{j-1} p_{J_k}(x) + p_{J_k}(j) p_{L_k}(j), & j > L_0 \\ 0, & o.w. \end{cases} \quad (8.21)
$$

where $\epsilon_J^k = 1 - \sum_{x=0}^{L_0} p_{J_k}(x)$ and $\epsilon_L^k = 1 - \sum_{y=0}^{L_0} p_{L_k}(y)$. Given the pmf of the real GDI functions, the upper bound of $I(R; A)$ is calculated from (8.18)-(8.21).

Since the observed rekeying process is determined by the artificial GDI, and the artificial GDI is only related with the real GDI, the following Markov chain can be formed: real GDI $\rightarrow$ artificial GDI $\rightarrow$ observed rekeying process. Thus, the mutual information between the observed process and the real GDI is no more than the mutual information between the real and artificial GDI [30]. Therefore, $I(R; A)$ is the upper bound of the amount of information that can be possibly revealed from the observed rekeying process.

From (8.12)-(8.14), one can see that the artificial GDI reveals the real GDI when $N(k) > N_0$, $L(k) > L_0$, or $J(k) > L_0$. Therefore, another useful metric is the *overflow probability*, defined as the probability that the artificial GDI cannot be straight lines, i.e. $1 - \min_k(1 - \epsilon_N^k)(1 - \epsilon_L^k)(1 - \epsilon_J^k)$.

Besides the mutual information, overflow probability can be a complementary measure for the leakage of the GDI. When the overflow probability is zero, the calculation in (8.18)-(8.20) leads to the result that $I(R; A) = 0$, which indicates prefect protection of the real GDI.

8.3.2 Communication Overhead

Communication overhead, measured by the rekeying-message-size, is one of the major performance criteria of key management schemes [78] [7]. We introduce the notation $M(L, N, d)$ as the expected value of the rekeying-message-size when removing L users from the key tree that contains total N users and has degree d. We assume that the leaving users are uniformly distributed on a fully-loaded and balanced key tree. Then, there are d^l KEKs at the lth level of the key tree for $l = 0, \cdots, D-2$ and $D = \lceil \log_d N \rceil$, and the number of the KEKs at the $(D-l)^{th}$ level is $s_1 = \lceil \frac{N-d^{D-l}}{d-1} \rceil$.

Let α^l be the number of the KEKs need to be updated at the level l when L users leave the group. Then, $M(L, N, d)$ is

$$M(L, N, d) = E\left[\sum_{l=0}^{D-1} \alpha_l\right] = \sum_{l=0}^{D-1} E[\alpha_l]. \qquad (8.22)$$

The expectation $E(.)$ is taken over the statistics of user departure behavior and the dynamic tree structure.

In Chapter 5, we have introduced the notation $B(b, i, a)$, used to calculate rekeying message size. With this notation, $E[\alpha_l]$ is calculated as

$$E[\alpha_l] = d \cdot B(d^l, L, \frac{N}{d^l}), \quad 0 \leq l \leq D - 2, \qquad (8.23)$$

$$E[\alpha_{D-1}] = (d-1) \sum_{x=1}^{L} \binom{s_1}{x} \binom{N - s_1}{L - x} / \binom{N}{L} B(s_1, x, d). \qquad (8.24)$$

Using the fact that $\lceil \frac{i}{a} \rceil \leq B(b, i, a) \leq \min(b, i)$ (see Appendix), one can derive the upper bound of the $M(L, N, d)$ as:

$$M(L, N, d) \leq dL \log_d(N). \qquad (8.25)$$

This upper bound indicates that the communication overhead increases linearly with the number of departed users and with the logarithm of the group size.

Let C_r and C_a be the average communication overhead for rekey process based on real GDI and the artificial GDI, respectively. Then, the extra communication overhead introduced by the proposed defense technique is:

$$C_a - C_r = \frac{1}{T} \sum_{k=1}^{T} M(L_a(k), N_a(k), d) - \frac{1}{T} \sum_{k=1}^{T} M(L(k), N(k), d). \qquad (8.26)$$

When the overflow probability is small, (8.26) can be approximated by:

$$C_a - C_r \approx M(L_0, N_0, d) - \frac{1}{T} \sum_{k=1}^{T} M(L(k), N(k), d). \qquad (8.27)$$

8.3.3 System Optimization

From the system design points of view, parameter L_0 and N_0 should be chosen such that the leakage of the GDI is minimized while the extra communication overhead do not exceed certain requirements. When the overflow probability is small, the optimization problem is formulated as:

$$\min_{N_0, L_0} \sum_k H(N_a(k)) + \sum_k H(J_a(k)) \qquad (8.28)$$

subject to:

$$M(L_0, N_0, d) \le \beta, \qquad (8.29)$$

where β is the maximum allowed communication overhead per key updating. Additionally, $H(N_a(k))$ in (8.20) is monotonous non-increasing with N_0; $H(J_a(k))$ in (8.19) is monotonous non-increasing with L_0; and the communication overhead $M(L_0, N_0, d)$ in (8.22) is non-decreasing with L_0 and N_0. Therefore, the optimization problem is simplified as:

$$\min_{L_0} \left(\sum_k H(N_a(k)) + \sum_k H(J_a(k)) \right) \Bigg|_{N_0 = M^{-1}(\beta)|_{L_0,d}}, \qquad (8.30)$$

where $M^{-1}(\beta)|_{L_0,d}$ is the largest value of N_0 that satisfies (8.29) with given L_0 and d. Fortunately, the number of departed users between two key updates is usually much less than the group size. Thus, the search space for parameter L_0 is not large and this optimization problem can be solved by a full search.

8.4   Simulations

Mlisten [72], a tool developed at Georgia Institute of Technology, can collect the join/leave time for the multicast group members in MBone [90] sessions. The defense scheme described in Section 8.2 is applied to the data collected in 1996 [127]. Particularly, one audio session that started on Oct. 29th and lasted for about 5 days and 20 hours is selected. Figure 8.4 shows the values of $N(k)$, $L(k)$ and $J(k)$ of this session, where B_t is chosen to be 15 minutes.

It is suggested that the users' statistical behavior, such as inter-arrival and membership durations, can be modeled by exponential distribution in a short period of time [90]. In the simulation, the entire service time is divided into non-overlapped sections, as illustrated in Figure 8.4. The

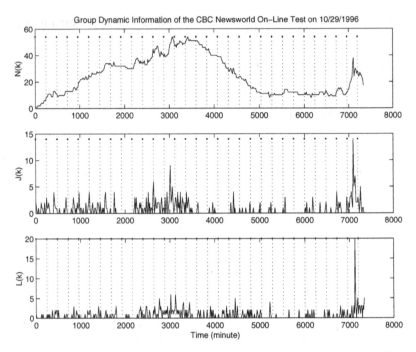

FIGURE 8.4. The GDI of a long audio session in MBone

length of these sessions is set to be 4 hours. To simplify the analysis, it is assumed that $N(k)$, $L(k)$ and $J(k)$ are stationary and ergodic Poisson processes in each session. Then, we can calculate the GDI leakage using (8.18)-(8.21).

Figure 8.5 and Figure 8.6 demonstrate the upper bound of mutual information (see (8.18)) and the communication overhead $M(L_0, N_0, d)$ for different values of L_0 and N_0, respectively. It is noted that these two figures use different axis in order to show the properties of the 3D curves. It is seen that communication overhead is a non-decreasing function with L_0 and N_0, while the GDI leakage is a non-increasing function with L_0 and N_0.

Figure 8.7 illustrates the solution of the optimization problem. Figure 8.7(a) shows the maximum value of N_0 that satisfies the communication overhead constraint in (8.29) with fixed L_0, i.e. $N_0 = max\{N : M(L_0, N, d) \leq \beta\}$, where β is chosen to be 50 in this example. As discussed in Section 8.3, the optimal values of L_0 and N_0 must be on this curve. Therefore, the upper bound of the GDI leakage, $\sum_k H(N_a(k)) + \sum_k H(J_a(k))$, is evaluated only at $(L_0, N_0 = max\{N : M(L_0, N, d) \leq \beta\})$, which is shown in Figure 8.7(b). The optimal values of L_0 and N_0 are also marked in Figure 8.7(b).

The tradeoff between the communication overhead and the GDI leakage is demonstrated in Figure 8.8. This figure shows the upper bound of the mutual information as a function of the communication overhead constraint, where the parameters L_0 and N_0 have been optimized. This can help the

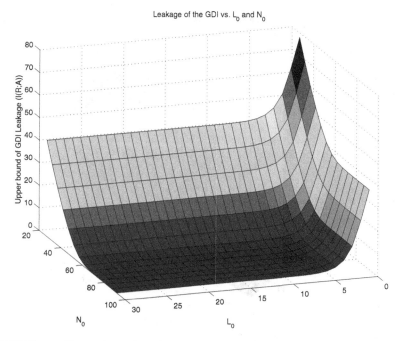

FIGURE 8.5. Upper bound of the GDI leakages (L_0 and N_0 are parameters in artificial GDI functions.)

system designer to determine the proper values of β for the communication constraint in (8.29). When not using the phantom users, the artificial process is identical to the real process and we have $I(R; A) = I(R; R) = H(R)$. In this case, this particular multicast session requires an average of 3.6 rekeying messages to be sent in every 15 minute interval ($B_t = 15$) and has $I(R; A) \approx 137$. Figure 8.8 shows that the proposed defense scheme can reduces $I(R; A)$ to 5.5 by increasing the communication overhead to 23.2 messages per 15 minutes. The communication overhead C_a is significantly larger than C_r because a large amount of activities of the phantom users must be created. However, the absolute value of the C_r is still small compared with the multicast data volume. On the other hand, the leakage of the group dynamic information is greatly reduced.

8.5 GDI Disclosure and Protection in Contributory Key Management Schemes

In the contributory key management schemes, every group participates the process of group key establishment. The members' personal keys are not disclosed to any other entities [56]. Compared with the centralized schemes,

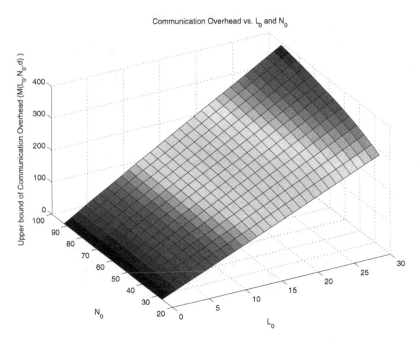

FIGURE 8.6. Communication overhead $M(L_0, N_0, d)$ (L_0 and N_0 are parameters in artificial GDI functions.)

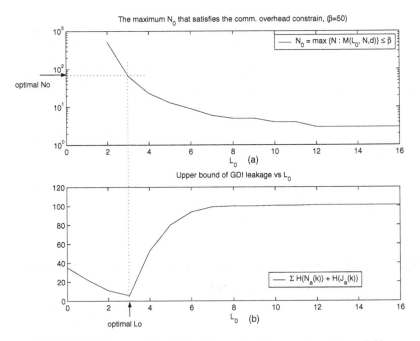

FIGURE 8.7. Illustration of selecting optimal parameters L_0 and N_0.

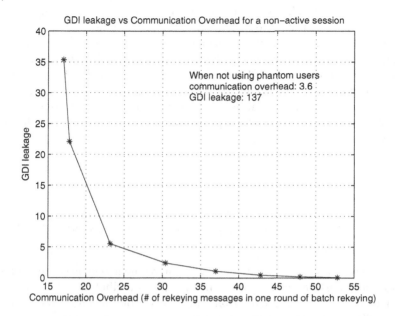

FIGURE 8.8. The GDI leakage versus communication overhead for a real MBone audio session, with and without phantom users.

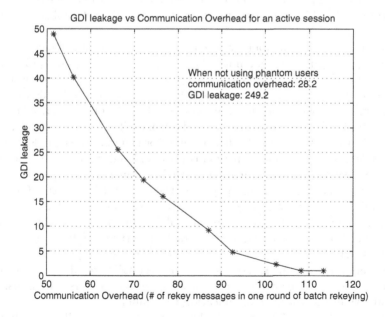

FIGURE 8.9. The GDI leakage versus communication overhead for a simulated multicast session, with and without phantom users.

the contributory schemes have the advantage of not putting full trust in a single entity and therefore do not suffer the problem of single-point-failure.

In general, the contributory schemes are suitable for small-medium group size applications, where group dynamics is known to group members. In these cases, protecting GDI is not necessary. On the other hand, it is possible that some special applications use contributory key management and require confidential GDI. In this section, we show that there are many ways to obtain GDI and the cost for protecting GDI in contributory schemes is very high.

8.5.1   Fully and Partially Contributory Key Management Schemes

There are two types of contributory key agreement schemes: fully contributory and partially contributory. In the fully contributory schemes, all key agreement operations are distributed to every group member [135]. There is no dedicated group manager, and every participant may perform admission control and other administrative functions [135]. Thus, group members are naturally aware of the information about the group membership. Therefore, the fully contributory schemes rely on the members' knowledge of dynamic group membership, and are not suitable for the multicast applications with confidential GDI.

In the partially contributory schemes, one group member takes a special role and performs some administrative operations [56, 58, 66, 67, 135]. This special member is usually referred to as the *group controller*. The role of the group controller can be assigned to a fixed member or be handed over to other members when membership changes [135]. The group controller is different from the KDC in the centralized schemes. The group controller does not hold the private keys of other members or generate the complete group key. Instead, it performs admission control and coordinates the process of the key formation. The original purpose of introducing a group controller is to achieve efficient key updating [56]. In the context of protecting GDI, the partially contributory schemes make it possible to confine dynamic membership information to the group controllers while preventing other group members from accessing GDI. In a practical setting, multiple group controllers, who are trusted to keep GDI, must be used to prevent the single point failure problem. In addition, to protect GDI, regular users cannot replace the group controllers even if all group controllers fail. Thus, the reliability of the group communication may suffer.

As a summary, GDI can only be protected in partially contributory schemes, at the expenses of utilizing trusted group controllers and the risk of communication failure.

8.5.2 GDI Disclosure in Contributory Key Management Schemes

Utilizing a group controller is not a complete solution to the GDI protection problem. There are many other opportunities for the insiders to acquire group dynamic information.

The scheme presented in [52] is the earliest attempted to extend two-party Diffie-Hellman protocol to group applications. This scheme, sometimes referred to as ING [58], arranges members in a logical ring and is executed in $(n-1)$ rounds, where n is the group size. Therefore, every member obtains the group size by simply counting the number of rounds that he performed.

Similarly, the schemes presented in [53] and [54], referred to as the STR and BD respectively, also reveal the group size. Here, each member receives the broadcast messages from all other members, and therefore must know the existence of other group members.

In [66] [67] [61], logical tree structures are introduced to manage the formation of the group keys. In these schemes, each member performs L rounds and holds L subgroup keys, where L is the depth of the key tree. Since L is proportional to the logarithm of the group size, the group members know at least the order of the group size.

Another important set of contributory key management schemes are GDH.1, GDH.2 and GDH.3 [56]. These schemes arrange group members in a logical chain and accumulate the keying materials by traversing group members one by one. In GDH.1/2, the kth member receives k or $k+1$ messages from the $(k-1)^{th}$ member. Thus, the number of the messages reveals information about the group size. The users who are closer to the end of the chain have more accurate information about the group size. GDH.3 is executed in four stages [56]. In the second and the fourth stage, the last user on the key chain broadcast n messages to the rest of the group, and n is the group size. In all three schemes, the group size information is revealed by the size of keying messages.

8.5.3 The Cost of Preventing GDI leakage

It is seen that hiding GDI in contributory schemes is a very difficult task. Therefore, we suggest using the centralized key management schemes for the applications with confidential GDI. However, if the centralized schemes cannot be employed and GDI must be protected, which is a very rare case, a possible solution is to use GDH.3 with two modifications. The first modification is to use the group controller. Among all contributory schemes, GDH.3 has the strongest centrality flavor. The group members are arranged in a logical chain, and the group member at the end of the chain takes more responsibility than other members. If the group member at the end of the logical chain is selected as the group controller, which performs admission control and coordinates the key formation, a regular member only needs to communication with his two neighbors on the key chain and the group controller. The second modification is to replace broadcast messages with multiple unicast messages. This is necessary to prevent GDI leakage through the size of the broadcast messages. In addition, anti-traffic-analysis techniques, such as those in [134] [136], should be used to prevent GDI leakage to the outsiders. This possible solution yields unbalanced load among group members and significantly increases protocol overhead and complexity. The high cost and complexity make the GDI protection not practical in contributory environment. As a summary, contributory key management is not suitable for applications requiring GDI protection. The centralized key management scheme should be used for the applications with confidential GDI."

8.5.4 More on GDI Leakage Problem

Key management is not the only source, but is a critical source of GDI leakage. Attacks based on key management are effective, stealthy, and easy to launch. An attack who registers as a group member or monitors rekeying traffic near a group member, can obtain a large amount of GDI information without being detected.

Besides key management, monitoring multicast data delivery is another dimension for acquiring GDI. For the purpose of debugging, management and modeling, various tools have been developed to monitoring multicast communications [73]. If the underlying multicast applications are "cooperative", i.e. not using any preventive methods, one can obtain GDI using these tools. Generally speaking, the attacks based on data delivery monitoring are less attractive than those based on key management for two reasons. First, encryption and anti-traffic analysis tools, such as onion-routing, can disable or significantly reduce the effectiveness of these monitoring tools. Second, these monitoring tools involve high implementation cost. For example, many require installing agents in multicast-enabled networks in order to collect data delivery or group information [73].

8.6   Chapter Summary

This chapter discussed the issues of the GDI disclosure in secure group communications. Particulary, it first presents several effective methods that could obtain dynamic group membership information from the current centralized key management schemes. Then, it presents defense techniques that could protect GDI, by utilizing batch rekeying and phantom users. The fundamental tradeoff between the communication overhead and the leakage of GDI was studied. Finally, a brief discussion on the GDI problem in contributory key management schemes was provided. It was argued that contributory schemes were not suitable for applications in which GDI should be protected.

9
Reducing Delay and Enhancing DoS Resistance in Multicast Authentication

One security service that has been difficult to provide for multicast is authentication. Existing solutions are either resource-intensive, or introduce significant delay in authentication. A consequence of the delay overhead associated with many multicast authentication schemes is that they rely on receiver-side buffers and are therefore susceptible to denial of service (DoS) attacks targeted at filling a receiver's buffer with false packets. Therefore authentication strategies that allow for less delay and more efficient utilization of buffer resources are desirable.

One explanation for the inefficiency associated with multicast authentication stems from the underlying conceptual formulation of authentication. Authentication is about trust, and in the context of traditional network security services, trust is a binary concept. A binary formulation of trust is a deviation from our natural, social understanding of trust where the confidence we place in others is not a black-and-white concept, but rather broken down into many shades of gray.

In this chapter, our objective is to present strategies that reduce the delay associated with multicast authentication, make more efficient usage of receiver-side buffers, make delayed key disclosure authentication more resilient to buffer overflow denial of service attacks, and allow for multiple levels of trust in authentication. Throughout this chapter, we will focus our discussion on the popular multicast authentication scheme, Timed Efficient Stream Loss Tolerant Authentication (TESLA), though our techniques can apply to other authentication methods based upon the delayed key disclosure principle. Like other schemes based upon delayed key disclosure,

TESLA is susceptible to DoS attacks and is not well-suited for delay-sensitive applications.

At the heart of our approach is a modification to TESLA, which we call *Staggered* TESLA, that employs several message authentication codes (MACs) that correspond to authentication keys that are staggered in time. Staggered MACs provide notions of partial authentication and allows for forged packets to be more readily removed from the buffer, thereby improving usage of the receiver's buffer. A benefit of partial authentication is that one may define security policies that allow for partially authenticated packets to pass through the buffer, and thus packets will remain in the buffer for a shorter duration. In many scenarios accepting partially authenticated packets is unacceptable, and therefore we present two further techniques that may be used to reduce the delay needed for full authentication. The first strategy requires that the source has a guarantee that there are no adversaries within a certain network distance of the source. By having a guarantee of proximity protection, partially authenticated packets may be accepted as fully authentic. The second strategy for reducing full authentication delay that we present involves replicating the key distribution functionality within the network, and having a set of distributed key distributors transmit the key seeds. A benefit of all of these strategies is that they mitigate the threat of a buffer overflow DoS attack since an adversary must conduct a DoS attack at a higher attack rate.

The rest of the chapter is organized as follows. In Section 9.1, we review the related works in multicast source authentication, and give a brief overview of the conventional TESLA scheme. We explore partial trust and use it in Section 9.2, where we describe the Staggered TESLA scheme. The security requirements needed to reduce full authentication delay will be discussed in Section 9.3. We derive theoretical guidelines for buffer requirements and discuss the tradeoffs involved in Staggered TESLA in Section 9.4. We support the theoretical analysis by conducting simulations and present the results of the simulations in Section 9.5. Finally, Section 9.6 concludes the chapter.

9.1   Background Literature and TESLA

9.1.1   Related Work

Source authentication enables receivers to verify that the received data originated from the claimed source and was not modified. Source authentication in point-to-point communications can be solved by asymmetric cryptography. Asymmetric cryptography, however, consumes significant communication and computational resources that cannot be supplied by resource-limited devices. Source authentication can also be accomplished through symmetric cryptography by appending MACs to each packet. The problem of authenticating multicast is more complex than the unicast case when there are untrusted receivers in the multicast group. Simply applying MACs does not provide source authentication in multicast. Adversarial group members, who share the same secret key as benign group members, can easily create packets with MACs using this shared key. Since all users share the same key, the receivers cannot resolve the source of the packets.

Although digital signatures [13] can be applied to multicast authentication, they have prohibitive computational and communication overhead. Gennaro [137] and Wong [138] proposed schemes to mitigate communication overhead by amortizing a single signature across several packets. Rohatgi [139] introduced an improved approach that employs k-time signature schemes and has less delay. Another signature amortization scheme is based on an information dispersal algorithm that can tolerate certain amount of packet loss [140, 141]. Recent efforts on signature amortization for multicast authentication have involved distillation codes and have focused on resistance to denial of service attacks [142]. Another work along these lines was presented by Lysyanskaya et. al. [143] in which a multicast authentication scheme based on a combination of digital signatures, hashes and error correction codes is presented.

Multicast source authentication based on symmetric cryptography has attracted intensive research. Canetti presented a solution to multicast source authentication based on verifying l different MACs using l different keys for each message [144]. Unlike the method proposed in this chapter, the multiple MACs in [144] are calculated using independent keys that are not temporally linked. Further, their protocol is based on the assumption that no coalition of w bad receivers can forge packets for a specific receiver, but fails in the presence of a coalition of more than w users. Perrig constructed a signature scheme using one-way functions without trapdoors for broadcast authentication [145]. Xu and Sandhu [146] proposed two hop by hop authentication schemes suitable for Internet multicasting that use the multicast tree and is immune to DoS attack. A consequence of their hop-by-hop assumption is that intermediate routers are required to be trusted and secure.

Another popular approach uses delayed key disclosure. Delayed key disclosure was first introduced by Cheung [147] to achieve authentication for link state routing, and was used in the Guy Fawkes protocol to provide non-repudiation in unicast communication in [148]. Chained Stream Authentication [149, 150] and FLAMeS [151] used similar ideas for source authentication in multicast. In delayed key disclosure the sender keeps the key secret for some intervals of time after sending the data. The receivers buffer the packets since they do not have the authentication key. A short time later, the sender discloses the key and the receivers are able to perform authentication. Using delayed key disclosure introduces two new issues. The first issue is the buffer requirements at the receiver. Because the receiver needs to buffer the received packets before it can authenticate them, an adversary can launch a DoS attack and fill up the receiver's buffer with bogus traffic. The receiver will have to drop packets due to a lack of buffer space. Second, many applications are sensitive to delay and reducing delay is critical for achieving desirable quality of service. As we shall discuss later in this chapter, reducing delay in delayed key disclosure schemes can be accomplished by either employing partial authentication or suitable assumptions about the application's security policy or the source's network neighborhood.

9.1.2   TESLA Overview

Among the many existing schemes employing the delayed disclosure principle, the TESLA [152–155] scheme is one of the most popular. We shall use the TESLA scheme as the basis for our discussions. TESLA is based on initial loose time synchronization between the sender and the receivers. TESLA divides time into intervals of equal duration, and each time slot n is assigned a corresponding key K_n. For each packet generated during time interval n, the sender appends a MAC that is created using the authentication key K_n. Each receiver buffers the packets, without being able to authenticate them, until the sender discloses the key K_n by broadcasting the corresponding key-seed s_n. Once s_n is disclosed, anyone with s_n can calculate K_n and can pretend to be the sender by forging MACs. Thus, the use of K_n for creating MACs is limited to time interval n, and future time intervals use future keys. Further, s_n is not disclosed until d time slots later, where d is governed by an estimate of the maximum network delay for all recipients.

The keys K_n are derived from s_n using a publicly available one-way function F', while the s_n are related to each other via a reverse-time chain of one-way functions. To create the chain of key-seeds, the sender chooses a terminal seed s_l, and generates s_{l-1} using a one-way function F. The remaining seeds $\{s_0, s_1, \cdots, s_l\}$ are derived via $s_l \xrightarrow{F} s_{l-1} \xrightarrow{F} s_{l-2} \xrightarrow{F} \dots \xrightarrow{F} s_1 \xrightarrow{F} s_0$. The sender uses the seed-chain in the opposite direction (starting with seed s_0) to derive the TESLA keys by applying the one-way function F' via $s_n \xrightarrow{F'} K_n$.

When a user receives a packet, he first checks whether the packet is *fresh* (i.e. it was sent in a timeslot whose TESLA-key has not been disclosed) or dated. The receiver discards all dated packets and buffers only the fresh ones. Once the user receives a TESLA-seed s_n, he checks $F(s_n) = s_{n-1}$ to be sure of s_n's authenticity. He derives K_n by $K_n = F'(s_n)$, and authenticates the packets that were sent in timeslot n. The conditions needed for the verification of the safe keys are collectively referred to as the security condition for TESLA. The use of chained key seeds also provides resilience to packet loss. If intermediate key seeds are not received, then a future key seed may be authenticated by applying the one-way function

F multiple times. The one-way function chain additionally allows for the determination of the packet's time of creation.

Several modifications are proposed in [153], where receiver buffering is traded-off at the expense of source buffering as well as a scheme, called concurrent TESLA, that is suitable for different receiver delays. The multiple MACs used in concurrent TESLA corresponds to multiple instantiations of the basic TESLA protocol, where each instantiation employs a different disclosure delay. This differs from the use of multiple MACs that we propose in Staggered TESLA, where our multiple MACs correspond to a single instance of TESLA using a single, assumed disclosure delay.

9.1.3 Examination of Trust in TESLA

We now examine the notion of trust in TESLA, and how it can be modified to achieve partial trust. In TESLA, the seed s_i for the authentication key K_i is released at a later time interval $i + d$, where d is a value greater than the maximum number of time intervals needed for a message to travel from the source to all of the receivers. As a result, the total time that a packet will occupy the receiver's buffer is approximately d intervals. Let us now consider the objective of the adversary. The adversary seeks to replace the content of the packets and make them pass the authentication check at the receiver. Thus the adversary needs to know the key K_i in order to successfully forge packets sent during interval i. Since the seed for key K_i is released during time interval $i + d$, the receivers do not accept any packets that claim to have been created during interval i after the start of time interval $i + d$. Thus adversaries are unable to forge MACs for interval i.

Now, let us consider what would happen if we send out the seed s_i earlier than in conventional TESLA. If s_i is sent at time interval $i+t$ instead of $i+d$, where $t < d$, then the receivers can authenticate packets sent in interval i when they receive the first packet sent in interval $i + t$. Consequently, the receivers can perform authentication sooner than they would have in conventional TESLA, and can thus remove the packets from the buffer earlier than in conventional TESLA. On the other hand, because the seeds are released earlier, some adversaries can take advantage of this and forge packets with valid MACs. Thus, authenticated packets cannot be classified as "fully trusted" and may be viewed as partially authenticated.

Our work is based upon this concept of partial authentication, and we therefore need to identify which entities are capable of forging packets with valid MACs at a specific time. Consider Fig. 9.1, where S corresponds to the sender, R depicts the receiver, and A_1 and A_2 are two adversaries at different network delay distances relative to the source and the receiver. A_1 is within a distance of $d - t$ from the source, while A_2 is a distance greater than $d - t$ from S. The distance in the figure represents the relative network time delay between entities for the transmission of a single key seed packet. These network delay positions might change from packet to packet

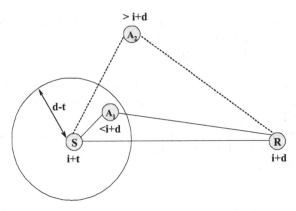

FIGURE 9.1. Network diagram depicting relative network distances for the source S, and receiver R for a single packet transmission. In TESLA, the network has a maximum network delay of d. For a single packet containing a key seed, adversary A_1 is within a radius of $d - t$ from the source, while adversary A_2 is beyond a radius $d - t$.

or interval to interval based upon network conditions. For simplicity, we assume that both adversaries do not require any time to process and forge packets. Additionally, we assume that the link between an adversary A_j and the receiver R is a very high-speed link (perhaps dedicated for the purpose of performing a DoS) and thus, for discussion, we consider the adversary-receiver links as 0-delay links. If the key is released in interval $i + t$, then all adversaries within a $(d - t)$ radius of the source, such as A_1, will receive the key before the start of interval $i + d$. Since the adversaries have 0-forge time and 0-delay links to the receiver, the receiver will receive packets forged by A_1 before the beginning of time interval $i + d$. The receiver will perceive that these packets obey the security condition, and put them into the buffer. Adversaries outside the circle of radius $(d - t)$, such as A_2, will receive the key after the start of interval $i + d$. Hence, they cannot forge packets with valid MACs. Therefore, exactly those adversaries that lie within a radius of $(d - t)$ delay from the source can successfully forge packets, and belong to the *forge-capable area* for that key seed. Hence, if we release the key seed at interval $i + t$, any packet from interval i that passes the authentication check can only be declared as partially trusted since there is a network area capable of forging that packet.

9.2 Staggered TESLA: Multi-Grade Multicast Authentication

We now use the idea of releasing key seeds earlier than in conventional TESLA to achieve multi-grade multicast authentication. We begin the same

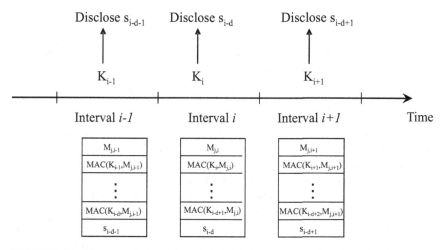

FIGURE 9.2. Format of the jth packet sent during interval $i - 1$, i and $i + 1$ in Staggered TESLA. There are d MACs attached to each packet, as compared to only 1 MAC attached to each packet in TESLA.

way as TESLA by splitting time into equal length intervals and assigning a seed s_n to time interval n. The authentication key K_n for interval n is derived from s_n using a publicly available one-way function F'. Our motivating observation is that many TESLA authentication keys will not be known by an adversary at an arbitrary location at an arbitrary time since it takes time for released keys to arrive at an adversary. Thus, when forging packets from interval i, which corresponds to key K_i being used to create MACs, adversaries might also not know K_{i-1} or K_{i-2}. Therefore, if we use more than just K_i to construct MACs during time interval i, such as using K_{i-1} and K_{i-2}, many potential adversaries will not be able to forge the MACs constructed using K_{i-1} or K_{i-2}, much less the MAC that used K_i. The idea of using MACs from successive TESLA keys leads to a scheme, which we call Staggered TESLA.

9.2.1 Format of the Packet

In TESLA, a MAC computed by the authentication key corresponding to the current interval is attached to each packet. Let $M_{j,i}$ denotes the jth message sent in interval i, K_i the authentication key used in interval i, and d the key disclosure delay in units of intervals. The source will disclose the key seed s_{i-d} in interval i. The receiver may use the seed to determine what time interval a packet was sent. The format of the jth packet sent in interval i is $\{M_{j,i}, MAC(K_i, M_{j,i}), s_{i-d}\}$.

In Staggered TESLA, we attach additional MACs made from previous TESLA keys to each packet. Because the seed s_{i-d} is released in interval i, attaching a MAC computed using key K_{i-d} is useless. Hence, the maximum number of MACs that can be attached in each packet is d, and instead of just attaching one MAC computed by K_i to each packet, we attach up to d MACs computed using $K_i, K_{i-1}, \cdots, K_{i-d+1}$, respectively. As shown in Fig. 9.2, the jth packet sent in interval i is

$$\{M_{j,i}, MAC(K_i, M_{j,i}), MAC(K_{i-1}, M_{j,i}), \cdots,$$
$$MAC(K_{i-d+1}, M_{j,i}), s_{i-d}\}. \tag{9.1}$$

Since Staggered TESLA uses consecutive, chained key seeds, it inherits the same resilience to packet loss as conventional TESLA.

We now discuss two issues related to the Staggered TESLA packet. First, we note that a simple and clever attack, which we shall call the *shift* attack, may be employed on the above packet format. In the shift attack, the adversary may take advantage of the fact that there is more than one MAC attached to each packet, and make use of the MACs from previous packets and shift them to forge later packets. For example, an adversary can store packet j from interval i, as in (9.1), and use it to forge the packets for interval $i + 1$ by sending

$$\{M_{j,i}, MAC(K'_{i+1}, M_{j,i}), MAC(K_i, M_{j,i}), \cdots,$$
$$MAC(K_{i-d+2}, M_{j,i}), s_{i-d+1}\}. \tag{9.2}$$

All of the MACs will be valid MACs except for the one using the fake K'_{i+1}, which the adversary could not forge. This attack, however, can easily be addressed by incorporating interval numbers and sequence numbers, as is typically done to prevent replay attacks [156], in the implementation when computing the MACs. Consequently, rather than complicate the notation in the remainder of the chapter, we stick with the above representation and note that the additional resources needed for appropriate indexing are minimal.

Second, the additional overhead for Staggered TESLA is minimal for many typical multicast scenarios. In particular, since MACs are based on symmetric cryptography, they are computationally efficient. Further, MACs produce short message digests, and therefore, the additional computation and communication requirements introduced by the extra MACs will not cause significant performance degradation. Consider a typical medium quality video multicast, where the average frame size is 1300 bytes [157]. In this case, the addition of a few 20 byte data fields, corresponding to a SHA-1 MAC, is minimal relative to the actual application data. We further note that one may employ fewer than d MACs, depending on the application's security requirements as well as the bandwidth restrictions of the underlying network, to reduce overhead.

9.2.2 Multi-Grade Source Authentication

In Staggered TESLA, the receiver-side buffer is a sequence of queues, as conceptually depicted in Fig. 9.3. When the receiver receives a packet, it

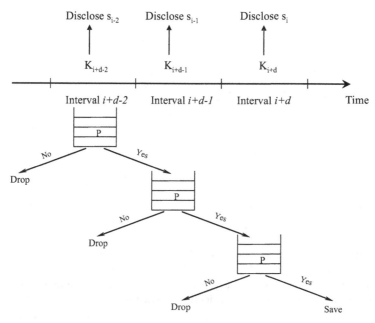

FIGURE 9.3. The events in Staggered TESLA and the chained buffer at the receiver. Partially authenticated packets graduate to lower layers of the buffer as the key seeds arrive at the receiver.

puts the packet into the top level of the queue, and graduates the packet to lower layers as additional key seeds arrive and the corresponding MACs are verified. If any verification fails, the packet is dropped from the queue, while if it passes, the packet becomes more trusted and graduates to the next layer of the buffer. This process repeats until the final key seed involved arrives and complete authentication is achieved. The chained buffer structure is easily implemented by tracking all packets waiting for a key, and updating which key each packet is waiting for following a partial authentication. Hence, the chained buffer we propose does not require any additional overhead compared with the traditional receiver buffer in TESLA.

In Staggered TESLA, if an adversary forges a packet for interval i, some of the MACs besides the normal TESLA MAC $MAC(K_i, M_{j,i})$ are likely to be wrong. Thus, most likely the receiver will be able to discard the forged packet before it would need to check $MAC(K_i, M_{j,i})$. Further, as a packet successively graduates from a higher layer to a lower layer in the buffer, the likelihood the packet is trustworthy increases. Thus, the receiver does not have to wait for the seed s_i in order to start authenticating packets. Instead, the receiver can use whatever seeds he/she has received to begin the authentication process and can promptly remove bogus packets. As a result, false packets are removed from the buffer quicker than in conventional TESLA. By contrast, in conventional TESLA, a forged packet will

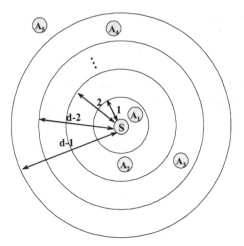

FIGURE 9.4. Adversaries at different locations pose different levels of threats. The receiver can remove false packets from more distant adversaries, such as A_5, sooner than those from closer adversaries, such as A_1. The forge-capable area shrinks as the packets pass authentication at each layer.

have to remain in the buffer for the complete disclosure delay before its falseness is revealed.

An individual packet that has had only some of its MACs verified is not fully authenticated and instead is only partially trustworthy. A packet's trustworthiness is directly related to which MACs have been verified and the amount of MACs employed in Staggered TESLA. Further, there is a direct relationship between a packet's position in the buffer and the size of the forge-capable area. Fig.9.4 shows the location of the sender and how the forge-capable area changes as key seeds are released. The distance in the figure denotes the relative time delay between the hosts, and for simplicity of discussion we consider that the network delay characteristics are fixed. Consider a packet that is sent during interval i, which has d MACs appended to it. These MACs are computed using keys $K_{i-d+1}, K_{i-d+2}, \cdots,$ K_{i-1}, K_i, respectively. Let's label these MACs as the $1st$, $2nd$, $\cdots$, and dth MAC. During time interval $i+1$, the seed s_{i-d+1} is released. The receiver is able to authenticate the $1st$ attached MAC after it receives the seed s_{i-d+1}. Since we assume the adversary-receiver link has delay 0, the forge-capable area for the $1st$ MAC is the circle of delay radius $d-1$ from the source. Adversary A_5, which is outside the circle, cannot forge packets with a valid $1st$ MAC. Thus, if there were an adversary at location A_5, the receiver would be able to remove all bogus packets sent by A_5 from the buffer at this time. However, adversaries within the radius $d-1$ circle, i.e A_1 to A_4, are able to forge the $1st$ MACs for any ith interval packet. Thus, the receiver cannot decide whether those packets are forged packets or not at this time.

At time interval $i + 2$, seed s_{i-d+2} is released. Now the receiver can perform authentication on the $2nd$ MACs, and similarly the forge-capable area shrinks to a region with radius $d-2$. Now both adversary A_5 and A_4 are outside the forge-capable area and both of them are unable to forge packets with valid $2nd$ MACs. The receiver can now remove all packets sent by adversary A_4. Similarly, the forge-capable area shrinks as the packets pass authentication at each layer of the buffer. There is progressively less area from which an adversary could successfully forge packets. Finally, during time interval $i + d$, the seed s_i is released, the forge-capable area has radius 0, and no adversary can forge packets with valid dth MACs. The receiver can fully authenticate the packet.

A packet gains trustworthiness as its forge-capable area shrinks. It is desirable to represent a packet's trustworthiness by a numerical value γ between 0 and 1. Such a quantification for partial authentication can allow for new security policies to be developed whereby partially authenticated packets are accepted if they have a threshold trust level. For example, a multimedia application might have strict QoS delay requirements and a security policy may be specified whereby, if the service quality provided to the user is not acceptable, the application would release additional packets whose γ is above a threshold set by the application designer.

The trust representation γ should be consistent across different Staggered TESLA sessions involving different interval sizes and different amounts of MACs. Hence, trust should be defined based on which MACs were used and which MACs were verified. Gambetta [158] defined trust to be the subjective probability that an agent can perform a particular action before that action can be monitored and before it affects a decision. There will be d or fewer MACs employed for each packet in Staggered TESLA, and a subset of these MACs will be verified. For Staggered TESLA, trust then corresponds to quantifying the likelihood that there are no adversaries that could have forged a particular subset of the MACs.

If we have a priori knowledge of the distribution for the delay τ between the source and a potential adversary, then we could take advantage of such information to define trust. Suppose that the adversarial delay τ has distribution f_τ. Then, for a Staggered TESLA scheme having d disclosure delay, which uses d MACs, and where the first t MACs have been verified, we may define trust as

$$\gamma(t) = 1 - \int_0^{d-t} f_\tau(\tau)d\tau \qquad \text{for } t < d. \qquad (9.3)$$

In the absence of any a priori distribution, two natural distributions that we may use are to choose τ to be uniformly distributed over $[0, d]$, or to assume that the network delay corresponds to propagation in two-dimensions and place the adversaries uniformly within a circle of radius d. This second choice is suitable for modeling delay in ad hoc networks, where a relationship between geographic location and network hop counts has been shown [159]. This leads to a distribution

$$f_\tau(\tau) = \begin{cases} 2\tau/d^2 & \text{for } 0 \leq \tau \leq d \\ 0 & \text{for } \tau > d \end{cases} . \tag{9.4}$$

In the case of the first distribution, our trust becomes $\gamma(t) = t/d$. On the other hand, in the second case the trust becomes $1 - (d-t)^2/d^2$. This definition of trust corresponds naturally our visualization of the forge-capable area as a circular region. In both cases $\gamma(d) = 1$, which corresponds to the trust level associated with full authentication of via the conventional TESLA MAC. We note that when measuring trust, we do not need to know the position of the adversary, but only which MACs have been verified.

To further illustrate the relationship between trust, network size and interval length, let us look at an example. Consider a network with a $400ms$ delay between the sender and the receiver, and an $800ms$ delay for the key release. If the interval size is $200ms$, then the key disclosure delay is 4 intervals. There are a total 5 levels of trust. However, if the interval size is $100ms$, the key disclosure delay will be 8 intervals, and there will be 9 levels of trust. There are tradeoffs between the selection of interval size and the number of levels of trust. If the interval size is large, there are fewer intervals and seeds needed. Hence, less communication overhead is needed to transmit those seeds. But at the same time, there are fewer levels of trust. On the other hand, there will be more levels of trust for smaller interval sizes. But this requires a longer key chain and larger communication overhead to distribute key seeds. Applications can select the interval size according to the network condition and security requirements.

The potential damage that can be caused to the authentication buffer is related to the adversary's location– the closer an adversary is to the source in terms of network proximity, the longer his forged packets will remain in the buffer. A coalition of adversaries may attempt a collusion attack, whereby the coalition shares key information with each other in order to facilitate an attack. We note, however, that in a collusion attack the adversary that is closest to the source is the most important member of the coalition as he is the one who will acquire the key seed first. Hence, even if there is a high-speed connection between adversaries, the strength of a collusion attack involving L adversaries is no greater than L times the strength of the closest adversary. Therefore, for the remainder of the chapter we shall only consider the case of a single adversary. The impact of the adversary's location on the receiver's buffer characteristics will be further discussed in Section 9.5.3. The end result is that Staggered TESLA allows the buffer to be more efficiently utilized and provides an advantage against DoS buffer overflow attacks. We will explore this behavior further in Section 9.4 and Section 9.5.

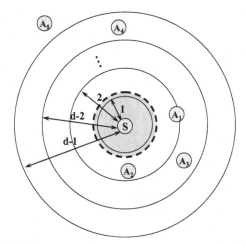

FIGURE 9.5. If there are guarantees that there are no adversaries located within the dashed circle of the source, packets can be fully authenticated one interval earlier in Staggered TESLA than in conventional TESLA.

9.3 Reduced-Delay Multicast Authentication Schemes

In the previous section, we discussed how partially authenticated packets can be released without waiting for the full authentication delay. We now examine two strategies that reduce the average delay needed for full authentication. The first scheme requires the assumption that the source has a guarantee of the trustworthiness of nearby network entities, while the second approach involves the introduction of additional key distributors, which are synchronized with the source.

9.3.1 Staggered TESLA with Proximity Protection

Adversaries at different locations pose different threat levels. The notion of forge-capable areas suggests that complete authentication is possible if we combine partial authentication with complementary forms of information assurance. One possibility, which we refer to as *proximity protection*, involves a guarantee that adversaries are not located nearby the source in network space. Proximity protection allows us to reduce the full authentication delay since, a few time intervals after the receipt of a Staggered TESLA packet, partial authentication will have reduced the forge-capable area to a small enough region that proximity protection will provide the remaining guarantee.

Consider, in Fig. 9.5, a source with proximity protection around his neighborhood so that it is guaranteed that there are no adversaries within the dashed circle. During interval $i + d - 1$, where key K_{i-1} is released, the forge-capable area shrinks to the region within radius 1, which is included in the dashed circle. No adversaries can forge the MACs computed using K_{i-1} successfully since they are all outside the forge-capable area of key K_{i-1}. Even though there is still one MAC left to be authenticated for each packet from interval i, the receiver can conclude that all packets that passed the authentic check for key K_{i-1} are actually fully authenticated and release those packets now. In order to save communication overhead, the source even does not need to attach the MACs corresponding to key K_i for packets from interval i. The larger the area the source can protect, the better we can reduce the full authentication delay. The amount of time intervals that can be reduced for full authentication corresponds to the largest forge-capable area within the protected region.

When specifying a region in network space near the source that can be protected, it is necessary to realize that the network delay will vary from packet to packet in a real network. Consequently, it is important to choose the region that the source can protect based upon the *minimal* delay that would be experienced by a packet traversing a portion of the network. In particular, since the queuing component of network delay might be 0, the region that can be protected should be decided based upon the non-variable components composing network delay, i.e. propagation, transmission and minimum processing delay. Another way to look at this is that the protected area must be the intersection of the guaranteed areas for all packets.

In practice, proximity protection can be realized in different ways for different types of networks. The relationship between network delay and hop counts suggests that the source needs to have a guarantee that network entities within a certain hop count are trustworthy. For networks like the Internet, this corresponds to having a guarantee that device's within the same access network are trustworthy. For networks like wireless ad hoc networks, hop counts can be related to actual physical distance from the source [159]. Thus, in such networks, network proximity protection becomes equivalent to physically guaranteeing that there are no adversaries within a geographic region around the source. Additionally, proximity protection may be achieved by appropriately employing scheduling and traffic control at intermediate routers (e.g. an overlay network or an ad hoc network) in order to ensure a specified level of minimum network delay.

Finally, we note that the authentication delay that can be reduced by proximity protection is quantized to multiples of the interval length. If an application needs to further reduce the authentication delay gap between the protected area and the largest forge-capable area, or the guaranteed area is too small to include any forge-capable area, it can use smaller intervals at the expense of additional communication overhead. Further, it might be necessary to shorten the interval length in order to have the resolution to define protected regions on networks where the non-variable delay component is small.

9.3.2 Distributed Key Distributors

We now present a scheme for reducing full authentication that may be used when there are no proximity guarantees. This scheme can be used with traditional TESLA or Staggered TESLA. For simplicity, we focus our discussion on applying distributed key distributors to TESLA.

We start by examining the total time that a packet will stay in the buffer in conventional TESLA, then we discuss the factors we can change to reduce delay. Consider a packet sent at time t_i in interval i, which takes l_i time units to arrive at the receiver. The first packet in interval $i+d$ that contains the key seed needed to authenticate the packets from interval i will be sent at time t_{i+d}. It takes l_{i+d} time units for this packet, and hence the key seed, to be delivered. Upon receiving this packet, the receiver can start authenticating packets from interval i. Thus, the total time that the packet from interval i will remain in the buffer is $t_{i+d} + l_{i+d} - t_i - l_i$. Among these four factors, t_i, l_i and t_{i+d} are unchangeable. We can control l_{i+d}, the time needed for the key to cross the network, by introducing additional key distributors in the network. These key distributors are trusted by the source, and possess a copy of the whole set of key seeds prior to the start of communication. The key distributors must be time-synchronized with the source, and will send out key seeds at the same pace as the source. Synchronization can be accomplished by employing standard methods, like NTP [160], to synchronize a set of distributed servers. The key distributors do not distribute content, but instead save communication overhead by sending only one key packet for each interval. The source can be thought as a special key distributor, which sends out keys and data at the same time. The use of the key distributors allows us to partition the network, where each network node belongs to the partition with minimum delay between the key distributor and itself. This reduces the average delay needed to receive the authentication key.

The key distributors can be placed at arbitrary locations in the network, though it is desirable to evenly place the key distributors in the network in order to better reduce the average authentication delay. If the network topology is known and the size of the network is small, the optimal locations can be obtained by exhaustive search. For larger networks, the k-means algorithm [161] can be modified to find the optimal locations for the key distributors. Each object is categorized in one of the k clusters according to the nearest neighbor policy. In our key distributor problem, the positions of the n network nodes yield k centroid points, where we place the k key distributors. However, it should be noted that a slight modification to the k-means algorithm is necessary since we do not have control over the position of the source, and must therefore fix one of the centroid positions, and determine the locations of remaining $k-1$ centroids in order to minimize network delay.

We now outline a modified k-means algorithm for placing the key distributors. As input to the algorithm, we assume that we have knowledge of the relative network delay positioning of each network entity.

1. Begin with an initial choice of k-1 nodes, together with the source as the centroid points.

2. Partition the whole set of objects into k clusters using the nearest neighbor policy.

3. Compute the centroid for each cluster and obtained a new set of centroid points, except the one with the source as it's centroid.

4. Compute the attribute for the new partition. If it has been changed by a small enough amount since the last iteration, then stop. Otherwise, go to step 2.

Just as in the traditional k-means algorithm, our modified k-means algorithm will converge to a local optima. In reality, network delay is variable, and the positions of the key distributors can be achieved using the estimated average delay.

Finally, we would like to briefly mention an alternative to distributed key distributors. The multiple key distributors are responsible just for transmitting key information. It is possible to replicate full multicast server functionality in the network and have the replicated servers transmit both content and key seeds. This has the effect of cutting the network into smaller networks for both the content distribution and the key distribution functions. Such a strategy is merely running multiple Staggered TESLA or TESLA servers, and therefore we will not consider it further in this chapter.

9.4   Buffer Requirements and Tradeoffs

When using Staggered TESLA, choosing an appropriate buffer size becomes an important issue. Too large a buffer size is a waste of resources, while too small a buffer will result in buffer overflow. In this section we revisit Staggered TESLA and explore the required buffer size for threat scenarios consisting of different adversarial attack rates. By explicitly calculating the average buffer size needed for the receivers, we provide guidelines for designing the buffer to fit the application and threat environment.

We employ a single adversary with the same network layout as depicted in Fig. 9.1. Let us consider $d + 1$ successive time intervals at the receiver. These intervals correspond to the receipt of packets sent in $d+1$ consecutive intervals, as represented in Fig. 9.6. We denote the duration of each time interval by T, which is a constant value. Throughout our discussion, we will assume that an adversary forges packets corresponding to the interval that

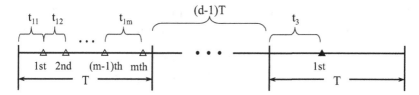

FIGURE 9.6. $d + 1$ consecutive intervals at the receiver. We depict the arrivals from the source during the first interval using blank triangles, and denote their interarrival times by t_{1j}. The 1st packet received in the $d + 1$th interval, whether from the source or the adversary, is depicted by the solid black triangle, and arrives after t_3 seconds after the start of the $d + 1$th interval.

is associated with the latest key seed the receiver knows. We will revisit this assumption in Section 9.5.

We will break the calculation of the buffer requirements into two parts: first, we will consider packets originating from the source, and then we will consider adversarial packets. After completing these analyses, we combine the two components to get the total average number of packets in the buffer.

For the first part, we assume that the packets sent by the source follow a Poisson process with parameter λ_1, and thus the interarrival times t_1 are governed by an exponential distribution with parameter λ_1, $p(t_1) = \lambda_1 e^{-\lambda_1 t_1}$. We assume that there are a total of m arrivals in the 1st interval that came from the source, and we denote their interarrival times as $t_{12}, \cdots,$ t_{1m}, as depicted by the blank triangles in Fig. 9.6. Since the interarrivals are exponentially distributed, we may use the memoryless property to define t_{11} to be the time from the start of the 1st interval to the $1st$ arrival, in which case t_{11} has the same distribution as t_1.

During the $d + 1$th time interval, the first packet that arrives, which we depict with a solid black triangle, may be either from the source or from the adversary. By the memoryless property of the exponential distribution, the time period from the boundary of the $d + 1$th interval to the arrival of the 1st received packet originating from the source in the $d + 1$th interval has the same distribution as t_1. Similarly, if the adversary emits packets as a Poisson process with parameter λ_2, then the time period t_2 from the boundary of the $d + 1$th interval to the 1st received packet originating from the adversary in the $d + 1$th interval has exponential distribution with parameter λ_2, $p(t_2) = \lambda_2 e^{-\lambda_2 t_2}$. Hence, the time period t_3 from the boundary of the $d + 1$th interval to the 1st received packet in the $d + 1$th interval is the minimum of t_1 and t_2, $t_3 = \min(t_1, t_2)$. Assuming that t_1 and t_2 are independent, then t_3 has exponential distribution with parameter $\lambda_1 + \lambda_2$, i.e. $p(t_3) = (\lambda_1 + \lambda_2)e^{-(\lambda_1 + \lambda_2)t_3}$.

Packets originating from the source during interval i can be authenticated when the receiver receives the first packet sent during interval $i + d$ because the packet contains the key seed needed to recover the authentication key K_i. In Fig. 9.6, all packets received in the 1st interval will be authenticated after the receiver receives the 1st packet in the $d + 1$th interval. Therefore, the total time W these packets will stay in the buffer are

$$
\begin{array}{ll}
1st & W = dT + t_3 - t_{11} \\
2nd & W = dT + t_3 - (t_{11} + t_{12})
\end{array}
$$

$$
\vdots \qquad \vdots
$$

$$
mth \qquad W = dT + t_3 - \sum_{i=1}^{m} t_{1i}. \tag{9.5}
$$

The expected value of W is

$$
E[W] = \sum_{m=0}^{\infty} E[W|M = m]p_m, \tag{9.6}
$$

where p_m is the probability of having m packets from time interval 1. The expected value of W conditioned on the total number of arrivals that originated from the source M is the average of the total time these M packets will stay in the buffer. Thus,

$$
E[W|M = m]
$$
$$
= \int \cdots \int \frac{mdT + mt_3 - \sum_{i=1}^{m}(m - i + 1)t_{1i}}{m}
$$
$$
p(t_{11}, \cdots, t_{1m}, t_3)dt_{11} \cdots dt_{1m}dt_3. \tag{9.7}
$$

Since $t_{11}, t_{12}, \cdots, t_{1m}$ are from independent exponential distributions with parameter λ_1, and t_3 has exponential distribution with parameter $\lambda_1 + \lambda_2$, the expected values are $\frac{1}{\lambda_1}$ and $\frac{1}{\lambda_1 + \lambda_2}$, respectively. Hence, (9.7) can be simplified as

$$
E[W|M = m]
$$
$$
= dT + \frac{1}{\lambda_1 + \lambda_2} - \frac{\sum_{i=1}^{m}(m - i + 1)}{m\lambda_1}
$$
$$
= dT + \frac{1}{\lambda_1 + \lambda_2} - \frac{(m + 1)}{2\lambda_1}. \tag{9.8}
$$

Substituting (9.8) into (9.6) and noting that M has Poisson distribution with parameter $\lambda_1 T$, yields

$$
\begin{aligned}
E[W] &= dT + \frac{1}{\lambda_1 + \lambda_2} - \frac{\lambda_1 T + 1}{2\lambda_1} \\
&= dT - \frac{T}{2} + \frac{1}{\lambda_1 + \lambda_2} - \frac{1}{2\lambda_1}. \tag{9.9}
\end{aligned}
$$

We will let $N_1(t)$ stand for the number of packets that originated from the source and are in the buffer at time t, while we will denote $\alpha(t)$ to be

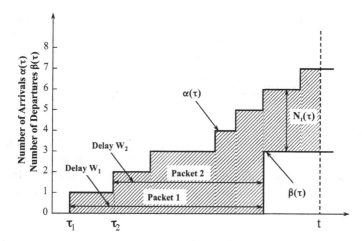

FIGURE 9.7. Arrival and departure process, $\alpha(t)$ and $\beta(t)$, for packets arriving at a receiver.

the number of packets originating from the source that the receiver receives and $\beta(t)$ to be the number of packets the receiver authenticates until time t. Further, let W_i be the time spent in the buffer by the ith received packet originating from the source. The time average of N_1 up to time t is

$$N_{1t} = \frac{1}{t} \int_0^t N_1(\tau)d\tau. \tag{9.10}$$

Usually N_{1t} changes with time t, but it tends to a steady-state value, N_1, as t increases, $N_1 = \lim_{t \to \infty} N_{1t}$. Similarly, the steady-state arrival rate of packets originating from the source during $[0, t]$ is defined as

$$\lambda_1 = \lim_{t \to \infty} \frac{\alpha(t)}{t}. \tag{9.11}$$

The average time in the buffer spent by a packet that originated from the source is

$$W = \lim_{t \to \infty} \frac{\sum_{i=1}^{\alpha(t)} W_i}{\alpha(t)}. \tag{9.12}$$

The arrival and removal of the packets sent by the source is shown in Fig. 9.7. The arrival process follows a single increase model, whereas the removal process is a multiple decrease model since many packets are flushed from the buffer simultaneously. From Little's Theorem [162], the relationship between N_1, λ_1 and W is $N_1 = \lambda_1 W$. Assuming ergodicity in the arrival process, we may equate the time average W with the ensemble average $E[W]$ to get $N_1 = \lambda_1 E[W]$.

The derivation of the average time to remove a false packet is similar to the above calculation. The receiver does not need to wait for the full disclosure delay period d to remove false packets in Staggered TESLA. Let d' be the number of intervals needed to remove a forged packet. Then the expected total time W' that a forged packet will stay in the buffer is

$$E[W'] = d'T - \frac{T}{2} + \frac{1}{\lambda_1 + \lambda_2} - \frac{1}{2\lambda_2}. \qquad (9.13)$$

From Little's Theorem, the average number of false packets N_2 can be expressed as $N_2 = \lambda_2 W'$. Again from the assumption of the ergodicity of the arrival process, we can equate W' with $E[W']$. Thus $N_2 = \lambda_2 E[W']$.

The packets in the buffer either originate from the source or from the adversary. Hence, the total average number of packets N in the buffer is the sum of those originating from the source and those from the adversary,

$$N = N_1 + N_2. \qquad (9.14)$$

We calculated the average number of packets in the buffer for different attack rates and for varying amounts of MACs employed in Staggered TESLA. The values were calculated according to (9.14), and are presented in Table 9.1, where the interval length was 200ms, the delay disclosure was 4 intervals, and the mean interarrival time from the source was 40ms. The first line of the table corresponds to the average interarrival time of the adversary's packets in units of ms. An infinite adversarial interarrival time corresponds to no adversary. If we place the adversary at a distant location relative to the source, the single MAC case, which corresponds to conventional TESLA, has $d' = 4$ intervals. Similarly, in this scenario, when we use 4 MACs, the number of intervals needed to purge forged packets is $d' = 1$. From this table, we see the advantage that Staggered TESLA provides as we increase the attack rate.

One way to think of the system is as an $M/G/\infty$ queue with two classes of arrivals, one from the source and the other from the adversary. These two arrivals are independent Poisson processes with different parameters and hence their sum is simply another Poisson process with parameter equal to the sum of the parameters of the two classes. These two classes of arrivals, however, have different service characteristics. Forged packets have a service time that depends on the availability of authentication keys, while non-forged packets must wait the full disclosure delay to be fully authenticated.

TABLE 9.1. (Theoretical) Average Number of Packets in Buffer

Rate(ms)	∞	40	20	10	5	2
#MACs= 1	18.0	35.0	52.5	87.5	157.5	367.5
#MACs= 2	18.0	30.0	42.5	67.5	117.5	267.5
#MACs= 3	18.0	25.0	32.5	47.5	77.5	167.5
#MACs= 4	18.0	20.0	22.5	27.5	37.5	67.5

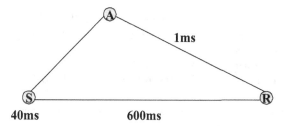

FIGURE 9.8. The arrangement of the source (S), receiver (R), and adversary (A) in the simulations. The connection between each pair of participants represents the aggregate link.

9.5 Simulations and Performance Analysis

We performed a series of event-driven simulations to evaluate the performance of Staggered TESLA, and techniques to reduce the full authentication delay. The first set of simulations, presented in Section 9.5.1 through Section 9.5.2, study the multi-grade property of Staggered TESLA. In these simulations, we assume there is no variability in the link delays. This allows us to deduce the effect of the adversary's network position on the buffering and authentication process. The second set of experiments, presented in Section 9.5.4, involves a more general network with variable delay links.

9.5.1 Simulations on Multi-Grade Authentication

The first set of experiments analyzes the improvement to the utilization of the receiver's buffer, as well as the impact that an adversary's location has on the buffer's behavior. We conducted two simulations that involved one sender, one receiver and one adversary, as shown in Fig. 9.8. In this setup, we abstracted the possible existence of multiple hops between each entity by representing the connection between each pair by a single, effective link. The first simulation we conducted is designed to show the effective usage of the receiver's buffer and the speed at which the receiver removes forged packets from the buffer when the adversary is at a fixed location. The second simulation shows how the performance of Staggered TESLA changes for specific locations of the adversary relative to the source and receiver. In both simulations, we collected statistics for the number of packets in the buffer and calculated the percentage of packets that actually originated from the source. Additionally, we recorded the total time needed to purge a forged packet from the buffer.

For both simulations, we set the length of a time interval to be $T = 200ms$, and the key disclosure delay to be 4 intervals. Both the source and the adversary send out packets as a Poisson process. The source sends out packets with an average of interarrival time of $40ms$, which is a typical sending rate for MPEG-4 video [157]. The average interarrival time of packets from the adversary is a parameter in the first simulation, and fixed at $5ms$ in the second simulation. The network delay between the source and the receiver was set to $600ms$. We assumed that the adversary has a fast link to the receiver with a delay of only $1ms$. The delay between the source and the adversary is set to $599ms$ in the first simulation, and varies in the second simulation.

The objective of a DoS attack is to keep the receiver's buffer as full as possible. We now look at the strategic issues governing the adversary's attack. An adversary has to decide which interval he will attempt to forge packets for before he transmits those packets. On one hand, an adversary does not want to send "old" packets that have already violated the TESLA security condition, as these will be immediately discarded by the receiver. On the other hand, if the adversary knows the key seeds before the receiver, he also does not want to release those seeds to the receiver because giving new key seeds to the receiver will help the receiver free the buffer even faster. Thirdly, the adversary also wants to make the bogus packets stay in the buffer as long as possible. Thus, the adversary should forge packets corresponding to the interval associated with the latest key seed that the receiver knows.

In order to reveal the behavior of Staggered TESLA under the worst possible threat scenarios, we empower the adversary by giving him knowledge of the difference between the sender-to-receiver network delay and the sender-to-adversary-to-receiver network delay. From the knowledge of the network delays, the adversary can figure out the newest key the receiver will know when he receives forged packets. The adversary should then transmit packets from the interval that corresponds to the release of that key. If the adversary has some of the keys to calculate the attached MACs, those MACs will pass the authentication check. If the adversary does not have the keys, he will fake those MACs with random bits. Those MACs will fail in the authentication check. The closer the adversary is to the source, the sooner he receives the key seeds and thus can attach more valid MACs to the packets, requiring a longer time for the receiver to remove the forged packets. In the worst case, if the adversary knows all the key seeds except the latest, it will take the receiver the full disclosure delay to flush bogus packets from the buffer.

9.5.2 Performance Analysis of Staggered TESLA

We now examine different sending rates for the adversary and the effect these rates have on the performance of Staggered TESLA. In order to gauge the efficiency of the staggered MACs to remove packets from the buffer, we set the buffer size to be sufficiently large so that buffer overflow does not occur for all adversarial transmission rates. We measured the number of packets in the receiver's buffer and calculated the proportion of packets in the buffer that originated from the source. Additionally, we computed the

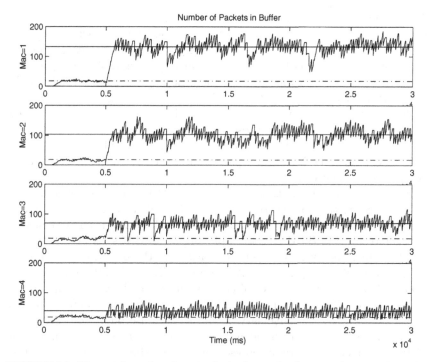

FIGURE 9.9. The number of packets in the buffer in the first simulation, where the Source-Adversary link has delay $599ms$. The four curves correspond to different amounts of MACs employed in Staggered TESLA. The adversary starts the DoS attack at 5 seconds.

total time needed to remove a false packet from the buffer. The simulation was run for $50s$, long enough for the system to achieve steady-state. We compared the performance for different amounts of MACs in Staggered TESLA. Since the key disclosure delay was 4 intervals in the simulation, the maximum number of MACs that could be employed in each packet was 4. Note that when only 1 MAC is attached to each packet, the situation is precisely conventional TESLA.

In the simulation, the adversary is set to be relatively far away from the source (at fixed delay of $599ms$). In order to demonstrate the behavior of Staggered TESLA during a normal traffic scenario and to illustrate the potential damage that our adversary can cause, the first 5 seconds involved only the source, and the adversary commences his denial of service attack after that.

Fig. 9.9 shows a realization of the number of packets in the buffer for the first 30 seconds, when the average interarrival time of packets from the adversary is $5ms$. Before the start of the adversary's DoS, the number of packets in the buffer is about 18. The number of packets in the buffer sharply increases as the adversary starts sending forged packets for all cases.

The dashed lines depict the average number of packets in the buffer before the start of the adversary's DoS, while the solid lines depict the average during the DoS. When the receiver receives a packet which does not contain a new key seed, the packet will be put into the buffer. When a packet is received that provides the key seed for a new key, all the packets in the buffer with MACs claiming to have been created using the new key will undergo authentication verification. If the adversary does not have the new key when he forges packets, those packets are proven false in this layer's authentication check and are then discarded. If the new key is the last key used to compute the MACs in a packet from the source, the packet will be completely authenticated.

It is clear from Fig.9.9 that when multiple MACs are employed in Staggered TESLA, the number of packets in the buffer is much lower during a DoS than in the case of conventional TESLA. A full-fledged version of Staggered TESLA (in this case employing all 4 MACs) is able to significantly reduce the average amount of packets in the buffer, even compared to the cases when only 2 or 3 MACs are used. In Table 9.2, we present the averaged values of the number of packets during the DoS for a time period of 45 seconds. The first line of Table 9.2 is the average interarrival time of packets sent by the adversary, in units of milliseconds. ∞ corresponds to no adversary, and can be identified with the first $5s$ of the simulation. Columns further to the right represent more powerful adversaries capable of conducting their DoS attack at higher attack rates. In all cases, the number of packets in the buffer increases as the adversary's sending rate increases.

The true advantage of Staggered TESLA is revealed when we examine the results within each column. For a fixed column, i.e. when the sending rate is fixed, the number of packets in the buffer is lower when there are more MACs in each packet. A more enlightening phenomenon is observed when we increase the attack rate of the adversary. For example, examining the columns for an attack rate of $40ms$ and $2ms$ (which corresponds to an increase in the attack rate by a factor of 20), we see that the number of packets in the buffer increases roughly by a factor of 10 for conventional TESLA, but only by a factor of 3 for Staggered TESLA with 4 MACs employed. Comparing the 4 cases in the table, full-fledged Staggered TESLA has the best performance.

Staggered TESLA not only decreases the number of packets in the buffer compared to conventional TESLA, but also improves buffer utilization

TABLE 9.2. Average Number of Packets in Buffer

Rate(ms)	∞	40	20	10	5	2
#MACs= 1	18.1	33.5	48.6	75.0	133.1	307.1
#MACs= 2	18.1	28.3	40.7	58.8	101.9	227.1
#MACs= 3	18.1	24.8	30.5	44.1	69.8	149.2
#MACs= 4	18.1	20.1	24.3	29.4	38.9	68.8

efficiency. In Table 9.3, we calculated the average percentage of packets in the buffer that originated from the source. In all cases, the utilization efficiency drops as the adversary's sending rate increases. For a fixed sending rate, the efficiency increases as we use more MACs, and shows the improvement that full-fledged Staggered TESLA provides compared to TESLA. Further, for full-fledged Staggered TESLA, the buffer utilization drops slower as we increase transmission rate than it does for conventional TESLA. Overall, these results mean that Staggered TESLA will provide improved resilience to buffer overflow attacks.

Overall, the use of multiple, staggered MACs in delayed key disclosure decreases the buffer requirements and more efficiently uses the buffer. These improvements are due to the fact that the receiver removes false packets faster in Staggered TESLA than in conventional TESLA. This can be explicitly seen in Table 9.4, where we present the average time needed to remove a false packet from the buffer. When there is only 1 MAC attached to each packet, it takes the receiver the full delay disclosure time to remove false packets. Since the key disclosure delay is 4, it takes the receiver 4 intervals to remove a false packet. Because some packets arrive earlier in an interval and some arrive later, the average time to remove a false packet is around $720ms$. When there are 2 MACs in each packet, it takes the receiver 3 intervals to remove a false packet and the average time to flush false packets is around $520ms$. The decrease in time is due to the fact that the adversary does not have both K_i and K_{i-1} when forging packets. When the number of MACs increases to 3, the number of intervals needed to remove a false packet decreases to 2. Finally, when there are 4 MACs appended to each packet, it only takes the receiver 1 interval to remove forged packets, which yields an average buffer time of around $120ms$. In this case, when the adversary forges the packets, he does not know any of the keys used to compute the MACs, i.e. K_i, K_{i-1}, K_{i-2} and K_{i-3}.

9.5.3 Impact of the Locations of Adversaries

We conclude from above discussion that the source should attach d MACs in each packet to optimize the performance when the adversary is "relatively" far away. We now examine the relationship that the adversary's position has upon Staggered TESLA. We emphasize that the advantages provided by Staggered TESLA do not depend on the ability of either the source or

TABLE 9.3. Average Percentage of Trusted Packets in Buffer

Rate(ms)	∞	40	20	10	5	2
#MACs= 1	1	0.56	0.40	0.24	0.14	0.06
#MACs= 2	1	0.64	0.47	0.32	0.19	0.08
#MACs= 3	1	0.74	0.60	0.43	0.28	0.13
#MACs= 4	1	0.88	0.80	0.68	0.55	0.34

the receiver to locate the adversary's relative position, nor does it depend on the ability to formally map out a forge-capable region in the network. Rather the performance advantages follow strictly from the use of multiple MACs.

The second simulation was conducted to analyze the effect of the adversary's position. The adversary's DoS behavior was fixed throughout all simulations as a Poisson source with an average sending rate of $5ms$ delay between consecutive packets. The adversary was placed at different, constant network delay distances from the source. We measured the number of packets in the buffer for different locations for the adversary. The average number of packets in the buffer is shown in Table 9.5. The first row in the table is the source-to-adversary-to-receiver delay in units of milliseconds. It was assumed that the adversary had a fast connection with which to attack the receiver, and thus the adversary-receiver delay was fixed to $1ms$. Thus, the adversary becomes progressively closer to the source as we move from left to right on the table. The ∞ delay corresponds to no adversaries.

From this table, we see that for scenarios where the adversary is further away from the source, there is improved buffer behavior as we use more MACs. On the other hand, when the adversary is closer to the source, there is little advantage to employing multiple MACs as the number of packets for different amounts of MACs is practically the same. A second observation that can be made from this table is that the number of packets in the buffer for different MACs can be divided into an amount of clusters that is roughly $1 + Delay/T$. For example, in the case where $Delay = 1ms$, there is a single cluster centered at 150 packets, while for $Delay = 400ms$ there appears to be three clusters: one at 157.5, one at 116.8, and one centered around 75. Similar observations can be made when one examines the average time needed to purge a forged packet, as presented in Table 9.6.

As we discussed earlier, keys sent out at different times will result in different forge-capable areas. The position of the adversary determines how many valid MACs he can forge when he sends out the forged packets. Let us consider a Staggered TESLA packet, sent during interval i, that consists of 4 MACs, such as

$$P_i = \{M_{j,i}, MAC(K_i, M_{j,i}), MAC(K_{i-1}, M_{j,i}), \cdots,$$
$$MAC(K_{i-3}, M_{j,i}), s_{i-4}\}.$$

TABLE 9.4. Average Time to Purge a Forged Packet

Rate(ms)	∞	40	20	10	5	2
#MACs= 1	N/A	724.5	723.5	726.4	724.1	725.5
#MACs= 2	N/A	522.8	521.2	524.0	523.8	522.7
#MACs= 3	N/A	327.5	325.8	327.4	326.9	323.7
#MACs= 4	N/A	125.1	124.7	124.3	121.9	125.8

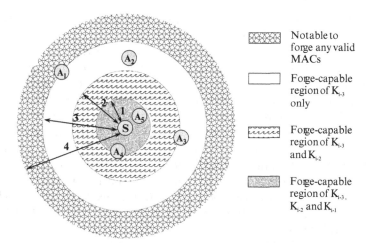

FIGURE 9.10. The position of different classes of adversaries, and their corresponding forge-capable areas for Staggered TESLA. Here, the key disclosure delay is $d = 4$.

As shown in Fig.9.10, when the path from the source to the receiver via adversary A_1 is $600ms$ (3 intervals), A_1 is just outside the forge-capable area of K_{i-3}. A_1 does not have any of the keys needed to compute the MACs when he forges the packets. Thus, when there are 4 MACs attached in the packets, the receiver can remove those forged packets from A_1 when he receives K_{i-3}. It takes the receiver only 1 interval to remove false packets. We have depicted the sequence of events leading to the removal of forged packets in this scenario in Fig.9.11 (a). During time interval $i + 3$, the adversary gets P_i and has K_{i-4}. At the same time, the receiver has received P_i and also has K_{i-4}. Recall that we assumed a powerful adversary who knows the state of the receiver he is attacking, and that the adversary will therefore forge packets corresponding to the interval associated with the latest key the receiver knows. Thus, during interval $i + 3$, the adversary will create forged packets P_i'. The adversary did not know any of the keys K_i, K_{i-1}, K_{i-2} or K_{i-3} and hence P_i' will only stay in the receiver's buffer for one interval. In contrast, if only 3 MACs were used in the packets, the receiver will only be able to remove false packets when he receives the seed for K_{i-2}, and the receiver must wait for 2 intervals before removing

TABLE 9.5. Average Number of Packets in Buffer

Delay(ms)	∞	600	500	400	200	1
#MACs= 1	18.1	133.1	156.4	157.5	158.0	157.2
#MACs= 2	18.1	101.9	115.6	116.8	118.7	148.4
#MAC= 3	18.1	69.8	78.6	77.9	109.3	146.9
#MACs= 4	18.1	38.9	53.2	71.4	110.8	150.6

false packets. At the extreme case, for conventional TESLA, it will take the receiver 4 intervals to remove false packets.

Consider adversary A_3, whose source-adversary-receiver delay is 2 intervals. He is outside the forge-capable area of K_{i-2}, but inside the forge-capable area of K_{i-3}. If the source attaches 4 MACs, then A_3 is able to forge the $4th$ MAC correctly. We present the sequence of events for A_3 in Fig. 9.11 (b). A_3 receives P_i during interval $i+2$, but will create a forged packet P_i' during interval $i+3$. Because A_3 knows K_{i-3} during time interval $i+3$, the receiver will not be able to reject P_i' until interval $i+5$, when it gets the seed to calculate K_{i-2}. Thus, it is actually $MAC(K_{i-2}, M_{j,i})$ that provides the ability to remove forged packets, and hence $MAC(K_{i-3}, M_{j,i})$ does not help to remove packets any faster. This explains why attaching 3 and 4 MACs gives roughly the same performance for an S-A-R delay of $400ms$ in Table 9.5.

Other locations for the adversary follow the same patterns. For adversary A_4, which is 1 interval away, attaching 2, 3 or 4 MACs gives roughly the same result. Finally, for adversary A_5, who can get the key seeds as soon as the source releases them, only one MAC is able to authenticate packets, and thus all cases give approximately the same result as TESLA.

An interesting observation can be made if we examine adversary A_2, who is 2.5 intervals from the source. A_2 is inside the forge-capable area of key K_{i-3}. At first glance, the case should be similar to the scenario for A_3 and have only 3 levels. However, since the adversary receives s_{i-3} in the middle of an interval, all packets forged before the adversary receives the seed will have wrong $4th$ MACs. For some adversarial packets, all 4 MACs are useful for removing packets, while for others only 3 MACs are useful. Thus, there are 4 levels for the number of packets in the $500ms$ column in Table 9.5, though the improvement of attaching 4 MACs over 3 MACs is not as large as for the $600ms$ delay case.

The theoretical values in Table 9.1 are close to the simulation results in Table 9.2 and 9.5. There are two factors affecting the differences. First, the theoretical calculations assumed the key seed is always available at the beginning of an interval for the adversary. In reality, this is not the case, and this effect is more pronounced when the adversary is far away. The adversary might receive P_i shortly into the interval, but until that time any forged P_i' will use an incorrect s_{i-d} and thus will be immediately dropped by the receiver since it will fail the key seed verification. This results in

TABLE 9.6. Average Time to Purge a Forged Packet

Delay(ms)	∞	600	500	400	200	1
#MACs= 1	N/A	724.1	707.0	704.8	705.0	704.8
#MACs= 2	N/A	523.8	507.2	504.3	504.8	664.4
#MACs= 3	N/A	326.9	307.2	303.9	461.8	663.0
#MACs= 4	N/A	121.9	177.0	262.8	464.4	666.7

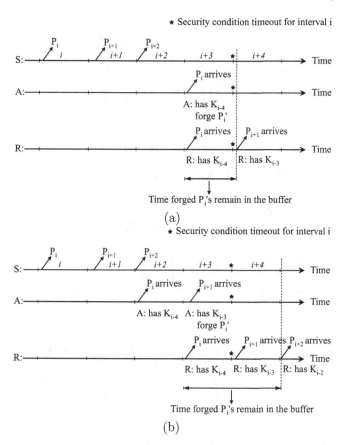

FIGURE 9.11. Sequence of events in Staggered TESLA. (a) The case when the Source (S)-Adversary(A)-Receiver(R) delay is $600ms$, (b) The case when the S-A-R delay is $400ms$.

the simulation having slightly lower values, as seen when comparing the $600ms$ cases with the values from Table 9.1. Second, the theory calculations assumed that, during each interval, the key seed is always available in the first packet that arrives at the receiver, though in actuality the first few packets might fail seed verification. Consequently, packets from interval $i - d$ remain in the buffer for a slightly longer period of time, and this effect can be seen in the case where 4 MACs are used. These two effects appear in different locations in the tables. Overall the discrepancies are small, which suggests that the theoretical calculations can serve as a good guideline for determining buffer requirements.

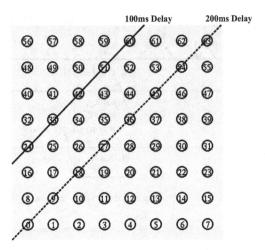

FIGURE 9.12. A grid network topology of 64 nodes. If the source can protect the region left to the solid line, then the average authentication delay can be reduced by 100ms. The dashed line is the assurance boundary of 200ms.

9.5.4 Simulation on Reducing Authentication Delay

The second set of experiments compare the full authentication delay and communication overhead of our two schemes described in Section 9.3. The simulations were conducted in the ns2 simulation environment with the network configuration shown in Fig. 9.12. The network is comprised of 64 stationary nodes located on an 8 × 8 grid. The distance between adjacent nodes is 50m, and the communication range and sensing range are set to be 50m and 100m, respectively. Thus, nodes can only communicate with their adjacent neighbors. Node 56 is set to be the source, which sends out packets as a Poisson source with 40ms average delay between successive packets. For simplicity, broadcast is used as the traffic dissemination pattern in the simulations. Each node only forwards newly received packets and discards all old ones. There is a fixed 25ms processing delay at each node before forwarding each packet. In addition to the variable queuing delay provided by ns-2, we added a random delay that was uniformly distributed between 0 and 10ms to reduce collisions. The payload of each packet is 1300 bytes, corresponding to a typical medium quality video [157]. We chose 802.11b for the ad hoc network, and thus the link bandwidth is 11Mbps. The farthest node, i.e. node 7, is 14 hops away via the shortest path and has a shortest path network delay of around 400ms. The addition of queueing delay and use of an alternative path may result in longer delay. The interval size is 100ms, and key disclosure delay is 800ms.

In Fig. 9.12, the 5-hop neighbors of the source will have delay greater than 100ms. If the source can employ proximity protection for the region to the left of the solid line (which denotes the 100ms delay line), then it is not necessary for the receivers to check the authenticity of the last MAC for full authentication. Further, it is not necessary to attach the last MAC in order to save communication overhead.

The positions of the key distributors are determined by the minimum sum of network delay according to the partition made by the key distributors. We placed the distributors by conducting a search to find the best locations for up to 4 key distributors. For the case of only 1 key distributor, either node 20 or 29 will be the solution. When 2 key distributors are placed, four combinations give the same result, nodes 10 and 38, nodes 11 and 45, nodes 11 and 46, nodes 18 and 38. Only one choice of locations is available for 3 key distributors, namely nodes 14, 18 and 45. Either nodes 10, 22, 34, 53 or nodes 13, 17, 43, 46 can be chosen for the 4 key distributors scenario. The source sends out packets as a Poisson process with 40ms average interrarival time, while the key distributors send out one key packet per interval.

The simulation results are shown in Table 9.7, where we compare TESLA, full-fledged Staggered TESLA, and Staggered TESLA with proximity protection of 100ms and 200ms delay, and the use of distributed key distributors with up to 4 key distributors placed in their optimal positions. The simulation was run for 50 seconds of network time, while in all cases steady-state was achieved in only a few seconds of simulation time. The five columns stand for the packet size in bytes, average authentication delay in ms, maximum authentication delay in ms, packet delivery ratio and the bandwidth consumed in bytes at each node for each data packet compared to TESLA. Each data packet consists of 24 bytes Physical Layer Convergence Protocol (PLCP) header and preamble, 24 bytes MAC header, 4 bytes Frame Control Sequence (FCS), 20 bytes IP header, 16 bytes released key, MACs (each of size 20 bytes), and the payload. For proximity protected delay of 100ms, only 7 MACs are attached to each packet, while only 6 MACs are attached to each packet for the proximity protection delay of 200ms. For key packets used by the key distributors, there is no payload and no MACs are attached, producing a key-bearing packet of size 88 bytes.

As presented in Table 9.7, both the proximity protection and distributed key distributor schemes can significantly reduce average authentication delay. The maximum authentication delay can also be reduced in most cases. Proximity protection for 100ms/200ms can reduce the average

TABLE 9.7. Comparison of Reduced-Delay Authentication Schemes

	Size	Avg	Max	Deliver	Bandwidth
TESLA	1408	780.33	1021.99	95.82%	100%
Staggered	1548	783.06	1050.44	95.39%	109.4%
100ms	1528	676.22	939.86	95.09%	107.7%
200ms	1508	580.64	842.08	95.35%	106.6%
Distributed1	1408	663.52	1009.58	95.19%	101.1%
Distributed2	1408	635.96	976.39	95.28%	101.3%
Distributed3	1408	616.61	954.54	95.61%	101.9%
Distributed4	1408	607.52	923.17	95.52%	102.0%

authentication delay by about 100ms/200ms. With only one key distributor added in the network, the average authentication delay will decrease by about 120ms, which is slightly better than proximity protection for 100ms. Adding extra key distributors will further decrease the average authentication delay. Due to collisions, in all cases the packet delivery ratio is about 95%. Compared to TESLA, there is an additional 9%, 7% and 6% communication overhead compared for full-fledged Staggered TESLA, Staggered TESLA with proximity protection of 100ms, and Staggered TESLA with proximity protection of 200ms. By comparison, there is only a 1-2% additional communication overhead for key distributors. It should be noted, however, that the reduced communication cost for distributed key distributors does not capture the overhead needed to maintain synchronized key distributors, or the cost needed to install the key distribution functionality at different locations in the network.

9.6   Conclusion

In this chapter, we have examined the concept of using multi-grade authentication, whereby varying degrees of trust can be incorporated into multicast authentication schemes based on delayed key disclosure. The Staggered TESLA multi-grade multicast authentication scheme was presented. Staggered TESLA employs multiple, staggered authentication keys that are used in creating the MACs for authenticating a packet. As a result, the receiver may partially authenticate a packet by using those authentication keys it has prior to the arrival of new key seeds. The use of these staggered MACs not only provides varying levels of authentication, but also reduces the delay needed to filter forged packets, thereby resulting in more efficient utilization of buffer resources compared to conventional TESLA. Theoretical results were provided that yield design guidelines for determining the appropriate buffer size. Further, theoretical and simulation results showed that the use of multiple MACs, and hence multiple grades of authentication, allows the receiver to flush forged packets quicker than conventional TESLA. As a result Staggered TESLA provides an advantage against a DoS attack as it requires an adversary to attempt a DoS at a higher attack data rate than is necessary in conventional TESLA. With the complementary forms of information assurance, Staggered TESLA can further reduce full authentication delay. We also examined a second strategy for reducing full authentication delay by introducing additional key distributors in the network. Simulations showed that Staggered TESLA with proximity protection, as well as the use of additional key distributors, is able to reduce authentication delay compared to TESLA, with a minor increase in communication requirements. In the next chapter we will examine an application of multicast authentication, where a delayed key disclosure multicast authentication scheme is used to establish a certificate and authentication framework for ad hoc and sensor networks.

10

An Authentication Service for Sensor and Ad Hoc Networks

10.1 Introduction

Remote sensing applications are becoming an increasingly important area for research and development due to the critical need for applications that will perform environmental monitoring, provide security assurance, assist in healthcare services and facilitate factory automation. In remote sensing scenarios, one or more applications are connected to a sensor network through a communication network. The sensors in the sensor network make measurements, such as local temperature or barometric pressure, and communicate this data with the appropriate application via the network. Providing security mechanisms for sensor networks is of critical importance since sensors will ultimately be used to assist in our daily lives. The authentication of the data source as well as the data are critical concerns since adversaries might attempt to capture sensors and tamper with sensor data. Traditional authentication frameworks based on public key cryptography are not suitable for sensor networks since the sensor network will ultimately consist of small, low-powered devices that are mobile. The limited computational and storage resources available to sensors necessitates alternatives to authentication based on public key certificates.

Recently, a set of *security protocols for sensor networks*, known as SPINS, has been proposed [163]. SPINS addresses authentication on limited resource sensor networks by introducing two security protocols that rely on the presence of a more powerful basestation and an initial shared secret between the basestation and each participating sensor node: SNEP and μTESLA. SNEP is a simple protocol that provides data confidentiality, two-party data authentication, and evidence of data freshness using only symmetric keys and counters. μTESLA is a modified version of the TESLA protocol, which performs bootstrapping without using a public key infrastructure (PKI) and discloses one key each *epoch* independently of the packet rate to provide broadcast authentication. Another work that focused on authentication for ad hoc networks was presented in [164]. In this chapter, a distributed light-weight model for authentication was presented that involves network nodes requesting trust references from neighboring nodes in order to establish the trust relationships needed for network authentication. Each entity maintains a list of trusted entities, and using these lists trusted communication paths between two arbitrary entities can be derived. One drawback of this method, however, is its scalability. For large networks, the size of the trust tables can become prohibitive. Another work on authentication for ad hoc networks that addressed the issue of scalability was presented in [165], which introduced the use of cluster heads to reduce the amount of control packets needed. In this work, the network is divided into cluster regions, and cluster heads are elected from the regular network nodes within each cluster. Authentication is provided by using a public key infrastructure that, unfortunately, is not suitable for small sensor devices.

These methods focus on ad hoc networks employing a flat topology. However, ad hoc networks have been recently shown to have capacity limitations, and one approach to address this drawback is to employ a hierarchical ad hoc network. In this chapter we will further explore the advantages of hierarchical ad hoc networks, particularly focusing on the advantages of the hierarchical ad hoc sensor network for performing authentication when compared with flat ad hoc networks. Authentication in hierarchical ad hoc networks has been essentially untouched, and we are aware of only one work in this direction [166], which focused on a military environment. The security of their work is based largely on the assumption that the access points, which corresponded to unmanned aerial vehicles, are unable to be compromised. This is an assumption that does not hold in non-military applications, and therefore we consider a three-tier hierarchical ad hoc network that is suitable for more general remote sensing applications running on the Internet. We develop an authentication framework for our three-tier hierarchical sensor network that addresses the hardware resources of the three-tier network, and employs cryptographic primitives that are appropriate for each type of node.

10.1.1 Hierarchical Sensor Network

Mobile ad hoc networking is the ideal architecture for the wireless sensor network since ad hoc networking provides a ubiquitous communication infrastructure capable of growing and adjusting to sensor dynamics. However, despite the popularity of flat ad hoc networks for sensor applications,

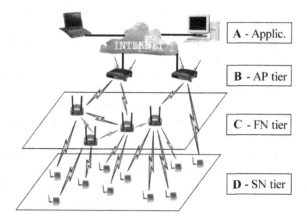

FIGURE 10.1. Three-tier hierarchical ad hoc sensor network.

recent information theoretic studies have indicated the limitations of the flat topology [167].

Recently, hierarchical ad hoc networks have been proposed as an alternative topology to flat ad hoc topologies. Initial measurements indicate that the hierarchical approach has better performance than flat ad hoc network [168, 169]. In [169], a three-tier self-organizing hierarchical ad hoc network is proposed to improve the scalability of ad hoc wireless networks. A modified hierarchical dynamic source routing (DSR) protocol [170] is studied. In particular, they observed, when using the same amount of sensor nodes in a given coverage area for flat and hierarchical topologies, that the system throughput capacity increases, while system delay decreases. The main reason for these improvements is the reduced number of hops since most sensor data are destined for the Internet, which is reachable in a few hops in the hierarchical approach.

In this chapter, we will use the three-tier ad hoc network topology of [169]. This architecture, depicted in Fig. 10.1, consists of three classes of wireless devices: (B) high-power access points that route packets received via radio links to the wired infrastructure, (C) mobile medium-powered forwarding nodes that relay information from sensor nodes to access points, and (D) low-powered mobile sensor nodes that have limited computing capability. We have depicted an Internet-based application (A) that is connected to the sensor network through the access points. This network is ideal for sensor-driven applications, where traffic flows from the sensors to Internet-based applications.

There are several key points that differentiate the three-tier hierarchical sensor network from conventional ad hoc sensor networks:

1. *Varying levels of computational power within the sensor network*: Conventional ad hoc sensor networks assume all nodes are created equal. The presence of large numbers of unreliable and energy-constrained sensors makes the task of energy-efficient communication and security protocols essential to the operation of a flat sensor network. However, the three-tier sensor network consists of three types of devices with different degrees of computational capabilities.

2. *Sensors do not communicate with each other*: The purpose behind a remote sensor network is to feed data to the application, which will make decisions based on the observations it receives. Sensor nodes route their packets via higher-level nodes and it is therefore unnecessary for sensor nodes to communicate and authenticate each other.

3. *The forwarding node is a radio-relay*: The purpose of the forwarding node is to relay messages from the sensors to the access points. Forwarding nodes have two wireless interfaces, one that communicates with SNs, and one that communicates with APs. They do not necessarily perform measurements themselves.

10.2   TESLA and TESLA Certificates

Today, the most widely used certification systems are PGP [171] and X.509 [15]. Both rely on public key cryptography, which makes them unsuitable for devices that are low-powered, or computationally-constrained. These devices should not have to verify an RSA-signature associated with a public key certificate. Therefore, if we wish to have a certificate-based authentication system for low-powered devices, we need a certificate structure that does not employ public key cryptography.

TESLA [152] is a broadcast authentication technique that achieves asymmetric properties, despite using purely symmetric cryptographic functions (namely MACs [172]) and thus enables low-powered nodes to perform source authentication. We now briefly review TESLA. TESLA divides time into intervals of equal duration. Time slot n is assigned a corresponding key tK_n. For each packet generated during time interval n, the sender appends a MAC that is created using the secret key tK_n. Each receiver buffers the packets, without being able to authenticate them, until the sender discloses the key tK_n by broadcasting the corresponding key-seed s_n. Once s_n is disclosed, anyone with s_n can calculate tK_n and can pretend to be the sender by forging MACs. Thus, the use of tK_n for creating MACs is limited to time interval n, and future time intervals use future keys. Further, s_n isn't disclosed until d time slots later, where d is governed by an estimate of the maximum network delay for all recipients.

The keys tK_n are derived from s_n using a publicly available one-way function F'. The s_n are related to each other via a reverse-time chain of one-way functions. To create the chain of key-seeds, the sender chooses a terminal seed s_l, and generates s_{l-1} using a one-way function F. The remaining seeds $\{s_0, s_1, \cdots, s_l\}$ are derived via $s_l \xrightarrow{F} s_{l-1} \xrightarrow{F} s_{l-2} \xrightarrow{F} \ldots \xrightarrow{F}$

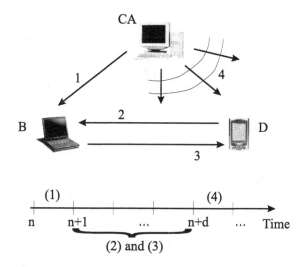

FIGURE 10.2. The steps involved in using TESLA certificates.

$s_1 \xrightarrow{F} s_0$. The sender uses the seed-chain in the opposite direction (starting with seed s_0) to derive the TESLA keys by applying the one-way function F' via $s_n \xrightarrow{F'} tK_n$.

When a user receives a packet, he first checks whether the packet is *fresh* (i.e. it was sent in a timeslot whose TESLA-key hasn't been disclosed) or dated. The receiver discards all dated packets and buffers only the fresh ones. Once the user receives a TESLA-seed s_n, he checks $F(s_n) = s_{n-1}$ to be sure of s_n's authenticity. He derives tK_n by $tK_n = F'(s_n)$, and authenticates the packets that were sent in timeslot n.

The framework that we propose in this chapter uses TESLA for authentication of data broadcast by the application. Additionally, TESLA is used to create TESLA certificates that support entity authentication in the sensor node handoff mechanism (cf. Section 10.7.2). In contrast to the original TESLA, our TESLA clients (i.e. all network nodes besides the forwarding nodes) are not bootstrapped using the public key infrastructure, but by the application sending the initial TESLA key to each network node encrypted with the appropriate shared key.

10.2.1 TESLA Certs

In Fig. 10.2, we present the entities involved in TESLA certificates, as well as the steps involved in using TESLA certificates. Much like conventional public key certificates, we have a certificate authority (CA), who is responsible for creating the certificates for entity B. A low-powered device, depicted by D, will contact B to use B's service.

The steps involved in TESLA certificates are:

1. The CA periodically issues TESLA certificates for B. During time slot n, the certificate authority (CA) doesn't sign the TESLA-certificate with its private key, but uses the non-disclosed TESLA key tK_{CA_n} to create a MAC that is included in the certificate. B's public key is replaced its authentication key aK_{B_n}, which is encrypted by the CA using the TESLA key tK_{CA_n}.

$$Cert_{CA_n}(B) =$$
$$(ID_B, \{aK_{B_n}\}_{tK_{CA_n}}, TS_A, MAC_{tK_{CA_n}}(...))$$

TS_A is a timestamp addressing the certificate's expiration date. The certificate is sent to B along with the matching authentication key aK_{B_n}:

$$CA \to B:$$
$$(Cert_{CA_n}(B), \{aK_{B_n}\}_{K_{CA,B}}, MAC_{K_{CA,B}}(...)).$$

2. Sometime between time n and $n+d$, D contacts B requesting to use B's service, $D \to B : (request)$.

3. Following the request in step 2, B must prove its identity to node D. B sends an authentication packet, which consists of the TESLA certificate and a MAC that was created using B's authentication key aK_{B_n}:

$$B \to D : (Cert_{CA_n}(B), MAC_{aK_{B_n}}(request)).$$

Upon receiving the authentication packet, D measures the freshness of the certificate by checking the timestamp of $Cert_{CA_n}(B)$ to make sure that it arrived before time $n + d$, when the CA announces tK_{CA_n}. If $Cert_{CA_n}(B)$ is fresh, D buffers the authentication packet.

4. The CA discloses its TESLA key tK_{CA_n} at time $n + d$. Upon receiving tK_{CA_n}, D checks the authenticity of the TESLA certificate by checking $MAC_{tK_{CA_n}}$, then it decrypts B's authentication key aK_{B_n} and checks $MAC_{aK_{B_n}}$. User D is able to certify the identity of B as long as it receives the TESLA certificate $Cert_{CA_n}(B)$ before the CA revealed the TESLA key TK_{CA_n}.

The lifetime of a TESLA-certificate is short. It depends on the disclosure time of the TESLA key that the certificate authority used when creating the MAC and encrypting the subject's authentication key. Choosing a key that will be disclosed soon lowers the delay in the authentication process at node D, but results in increased overhead when issuing new certificates.

10.3   Overview of the Authentication Framework

There are two primary goals for the framework: first, to ensure that the data received by the application (A) is sent by an approved sensor node (D); second, to verify that the data hasn't been modified on its way to the application. To achieve these goals, an authentication service has to be realized that authenticates incoming nodes, establishes shared secrets among them and with the application, keeps track of changes in the network topology and provides data origin authentication for sensor node data.

As in every authentication service, the proposed framework relies on the presence of initial trust. It is necessary to get some trustworthy information about an incoming node before it is allowed to join the network. Information that is provided by the node is not trustworthy if it is not confirmed by a trusted entity. Therefore, each node that wants to join the network must have a personal *initial certificate* (*iCert*) that is issued by the network's *trusted third party* (*TTP*). The *TTP* is nothing more than a reliable node that is able to perform RSA signatures, whose public key is known to all nodes of the network that are able to verify RSA-signatures.

When an access point B or a sensor node D wants to join the network, the node presents its *iCert*, which eventually will be checked by the application A (because of their role as a radio relay, we don't consider forwarding nodes here– their role in the network includes authentication upon request). If the *iCert* is valid, A will establish a shared secret $K_{A,B}$ (with B) or $K_{A,D}$ (with D), which will enable the new node to access the network and communicate with the application. Once the node is part of the network, its *iCert* becomes less important. As long as B or D doesn't switch applications, the trust relationship with the application will last. Instead, it needs a method to authenticate itself with other network entities that are not the application, e.g. an access point must be able to authenticate itself with a sensor node and vice versa. The topology of the ad hoc network may change frequently, and thus it is desirable for nodes to perform this inter-node authentication on their own, in a fast and flexible manner. The application A enables them to do so by periodically issuing *runtime certificates* (*cert*) for each access point and sensor node of the network. Since the computationally weak sensor nodes are included in this process, these *certs* can't rely on RSA signatures. As mentioned in Section 10.2.1 a new type of certificate will be used.

Having this network of authenticated nodes and shared secrets, we can use the trust relationships to provide a data origin authentication service. A sensor node D that wants to deliver data, creates a MAC using the secret key it shares with the application and appends it to the data. The application will use the MAC to verify the data's origin. D creates another MAC using the secret key it shares with its gateway access point B, which provides access to the Internet. B will use the MAC to perform access control, by making sure that D is a valid part of the network before forwarding D's data. In case one or more forwarding nodes are located between B and D, B will answer the challenge that D includes in the data-packet to assure that the data finally reaches B. While the challenge-response mechanism can guarantee that the data arrives at the access point, it is neither able to tell what happened to lost data nor to provide information about who dropped it. Since this method is unable to identify the entities involved in the data delivery, we refer to this mode of data delivery the *weak mode* of operation. However, if a sensor node wants to send sensitive data or, for any other reason, wants each node along the path to be authenticated before it actually starts sending the data, it can choose to use the *assured mode* of data delivery, which will provide authenticated information about those nodes at the cost of an overhead of message exchange and shared secrets.

In building our authentication framework, we assume each forwarding node and access point has an RSA-key-pair along with its certificate. We also assume that they know the TTP's public key $+K_{TTP}$. For reference, at the end of the chapter, we provide a summary of the notation we use in our framework.

10.4   Certificates

Certificates are the major tool to build an entity authentication service since they enable entities that don't share a secret key to establish a trust relationship. Our framework distinguishes between initial and runtime certificates. While each node needs an initial certificate to join the network, the runtime certificates are periodically issued by the application to during the network's lifetime. The forwarding nodes are an exception. A forwarding node needs a general certificate that is only used once a sensor node requests *assured service* (cf. Algorithm 18). In this case, an access point B checks the forwarding node's certificate. Therefore, certificates based on RSA can be used for the forwarding node's certificate.

10.4.1   Initial Certs

The framework relies on certificates as a means of *initial* trust. Each access point or sensor node that wants to join the network must own a certificate ($iCert_{TTP}$) issued by the network's trusted third party.

Access Point

We assume that the access point B is a device of high computational power and battery resources. This enables B to validate and perform RSA-signatures. Therefore, each access point has an RSA public and private key pair $(+K_A, -K_A)$ and an $X.509$-certificate ($iCert_{TTP}$) issued by the Trusted Third Party TTP, binding this keypair to their identity.

Sensor Node

The sensor node D applies to the trusted third party for its initial certificate ($iCert_{TTP}$). This initial certificate is tied to a certain application A that the sensor node plans to connect to.

$$iCert_{TTP}(D) =$$
$$(ID_D, \{iK_D\}_{+K_A}, TS_{TTP}, SIGN_{-K_{TTP}}(...))$$

The TTP issues this $iCert$ to the sensor node D along with the unique key iK_D. D uses this key to authenticate itself against application A. Since D's initialization key is encrypted with A's public key, only A is able to obtain the key from the certificate and proof A's authenticity. If D plans to connect to several applications D, it can apply for more than one certificate.

10.4.2 Runtime Certs

The purpose of runtime certificates is to maintain authenticity between the initial authenticated nodes during the networks's lifetime. As a result of the forwarding node and sensor node mobility, shared keys become obsolete and new keys have to be established. The runtime certificates use the trust relationships between the application and the nodes of the network to create new trust relationships among them.

Access Points

An access point runtime certificate must be readable by each sensor node D. Therefore, RSA-based certificates cannot be used. Instead, we will fall back on TESLA certs:

$$Cert_{A_n}(B) =$$
$$(ID_B, \{aK_{B_n}\}_{tK_{A_n}}, TS_A, MAC_{tK_{A_n}}(...)),$$

where aK_{B_n} is the access point's authentication key for slot n. B will use this certificate to authenticate itself with a sensor node during handoff. The $MAC_{tK_{A_n}}(...)$ proves that this certificate was issued by the application.

Sensor Nodes

The sensor node runtime certificates will be used by an access point B to check D's identity during handoff to establish a new shared secret. It contains D's ID, a timestamp TS_A and D's secret authentication key $aK_{AP,D}$ encrypted with an 'access point group key' gK_{AP}, that every access point of the network gets during its authentication with the application.

$$Cert_A(D) =$$
$$(ID_D, \{aK_{AP,D}\}_{gK_{AP}}, TS_A, SIGN_{-K_A}(...))$$

The signature $SIGN_{-K_A}(...)$ proves that this certificate is issued by the application.

10.5   Certificate Renewal

During the lifetime of a network, trust relationships change. Misbehaving nodes have to be identified and must not be allowed to remain connected to the network. Therefore, we need a certificate renewal mechanism.

10.5.1   Access Point

The application issues a new certificate for each connected access point B after a certain period of time. In the beginning of time slot n, A sends to B the new certificate:

$$A \to B : (Cert_{A_n}(B), \{aK_{B_n}\}_{K_{A,B}}, MAC_{K_{A,B}}(...)),$$

where aK_{B_n} is B's authentication key, which will prove B's identity to a sensor node D during handoff. Checking $MAC_{K_{A,B}}$, B can verify that this certificate was issued by the application A. To decrease the number of certificates that have to be issued, a certificate could contain several authentication keys, each of them encrypted with the TESLA key of a different time-slot, and the matching MACs at the end. However, in this chapter we concentrate on one key in each certificate.

10.5.2   Sensor Node

The application issues a new certificate for each connected sensor node D after a certain period of time. This period of time can be much greater than the TESLA time slots as the sensor node runtime certificates don't depend on the application's TESLA keys.

$$A \to D : (\{Cert_A(D), aK_{AP,D}\}_{K_{A,D}}, MAC_{K_{A,D}}(...)),$$

where $aK_{A,D}$ is the authentication key that D uses for sensor node handoff.

Algorithm: Access Point Authentication

Result: Authenticity and Shared Secret $K_{A,B}$
between Application A and Access Point B

1 $B \rightarrow A : (offer, SIGN_{-K_B}(offer), iCert_{TTP}(B))$

2 **if** *(A_accepts_offer)* **then**

    if *(SIGN_{-K_B}(offer) valid)* **then**

        $A \rightarrow B :$

        $(ok, SIGN_{-K_A}(ok), \{K_{A,B}, gK_{AP}\}_{+K_B})$

    else

        $A \rightarrow B : (deny, SIGN_{-K_A}(deny))$

    end

else

    $A \rightarrow B : (LoI, SIGN_{-K_A}(LoI))$

end

Algorithm 14: Access Point Authentication Algorithm

10.6 Entity Authentication

10.6.1 Access Point

An access point B isn't a proper mobile device, as it features a wired connection to the Internet and is installed at a certain place for a certain purpose. In the case of a sensor network the purpose is to provide Internet-access for sensor nodes and with that a connection to their application. There is a need for authentication of the access point because it will provide access control at the interface between the application and the sensors. Once an access point B is physically connected to the wired network, it will contact its application A and send a service-offer. We assume that B knows the address of A and A's public key $+K_A$. B will sign the offer with its private key $-K_B$ and append its certificate before sending the offer to the application A (cf. Algorithm 14). If A accepts the offer, it will check the signature with help of the certificate - otherwise it sends a signed *lack_of_interest(LoI)*-message. If the validation is successful it returns an accept-message including a shared secret key $K_{A,B}$ and the 'access point group key' gK_{AP} (that will be used in the sensor node handoff-scenario), encrypted with the access point's public key $+K_B$ and signed with its private key $-K_A$.

The *application $\leftrightarrow$ accesspoint* authenticity as well as the shared secret key $K_{A,B}$ are the basis for authenticity of the entire network.

10.6.2 Forwarding Nodes

Forwarding nodes are mobile devices. They don't feature wired connections and are free to roam between different access points or entire networks. In contrast to an access point, a mobile node needs a more flexible

---

Algorithm:*Sensor Node-Authentication at FN*

Result*: Authenticity+Shared Secret $K_{B,D}$ between Access Point B
and Sensor Node D;*
*Authenticity+Shared Secret $K_{A,D}$ between Application A and Sensor
Node D;*

1 $D \to C : (snReq, MAC_{iK_D}(snReq), iCert_{TTP}(D))$

2 $C \to B : (snReq, MAC_{iK_D}(snReq), iCert_{TTP}(D))$

3 $B \to A : (snReq, MAC_{iK_D}(snReq), iCert_{TTP}(D),$
$\qquad\qquad ID_B, MAC_{K_{A,B}}(snReq))$

4 **if** *($MAC_{iK_D}(snReq)$ valid)* **then**
$\qquad A \to B : (ok, MAC_{K_{A,B}}(ok), \{K_{B,D}\}_{K_{A,B}},$
$\qquad\qquad\qquad MAC_{iK_D}(ok), \{K_{A,D}, K_{B,D}\}_{iK_D})$
$\qquad B \to C : (ok, MAC_{K_{B,D}}(ok),$
$\qquad\qquad\qquad MAC_{iK_D}(ok), \{K_{A,D}, K_{B,D}\}_{iK_D})$
$\qquad C \to D : (ok, MAC_{K_{B,D}}(ok),$
$\qquad\qquad\qquad MAC_{iK_D}(ok), \{K_{A,D}, K_{B,D}\}_{iK_D})$
else
$\qquad A \to B : (nok, MAC_{iK_D}(nok),$
$\qquad\qquad\qquad MAC_{K_{A,B}}(nok))$
$\qquad B \to C : (nok, MAC_{iK_D}(nok))$
$\qquad C \to D : (nok, MAC_{iK_D}(nok))$
end

Algorithm 15*: Sensor Node Authentication Algorithm*

authentication mechanism to support its mobility. The forwarding node is
the only kind of device in the sensor network with two wireless network
interfaces. Its task is to forward data packets sent by sensor nodes, that
can't reach the access point directly or that can save energy by using the
forwarding node as an intermediate hop.

In the current approach, forwarding nodes only authenticate themselves
if a sensor node wants to send its data in the *assured mode* (cf. Algorithm
18).

10.6.3   Sensor Nodes

Sensor nodes are devices of high mobility with restricted computational
power. Our goal is to provide a flexible authentication scheme that supports
their mobility and relieves the sensors from intensive computations.

Once a sensor node enters the network, it sends its request to the application (cf. Algorithm 15). An intermediate forwarding node C simply forwards the request. An access point B that receives an sensor node

---

Algorithm:*Sensor node handoff*

Result*: Authenticity and a new shared secret $K_{B',D}$ between the senor node D and the new access point B'*

1 $B' \rightarrow D : (apHO, Cert_{A_n}(B'), MAC_{aK_{B'_n}}(...))$

2 $D \rightarrow B' : (apHO, Cert_A(D), MAC_{aK_{AP,D}}(...))$

3 $B' \rightarrow D : (hoOK, \{K_{B',D}\}_{aK_{AP,D}}, MAC_{aK_{B'_n}}(...))$

Algorithm 16*: Sensor Node Handoff*

---

authentication request appends its ID_B and a MAC that it creates using the key it shares with the application A. Once A gets the request, it checks the certificate. If the certificate is valid, A establishes a shared secret between the access point B and the sensor node D by returning two instances of a new key $K_{B,D}$, one encrypted with D's initialization key iK_D and the other encrypted with the key it shares with the access point. A also establishes a shared secret with the sensor node by appending another new key $K_{A,D}$, which it also encrypts using iK_D. Therefore, after receiving A's answer, D shares a secret with its gateway access point B and the application A.

10.7   Roaming and Handoff

The goal of "roaming" is seamless connection switching. In the hierarchical sensor network scenario, sensor and forwarding nodes switch between access points while moving or as a result of load balancing between access points. Roaming is a challenge for authentication mechanisms as trust-relations can't be reused since the network topology is changing quickly. In this section, we address authenticity problems associated with a forwarding node or sensor node changing their access point.

10.7.1 Forwarding Nodes

A forwarding node C doesn't connect to an application A or an access point B. Since C is not involved in any authentication processes, there are no shared secrets to update when C leaves the area near an access point. Thus, a forwarding node never has to perform handoff.

10.7.2 Sensor Nodes

During the lifetime of the network, the sensor node D will continually send data via the access point B to the application A. When the topology of the network changes in a way that D loses its connection to B and must connect to a new access point B', the data will no longer be delivered to A but will be blocked by B'. B' will then start the handoff-process by sending an *access_point_handoff_request* (*apHO*)-message to the sensor node (cf. Algorithm 16). Appended to the message is B''s TESLA certificate and a MAC that will be used by D to verify the identity of B' after A reveals the TESLA key. Next, D answers with its own certificate and a MAC that it creates using its authentication key $aK_{AP,D}$. B' will then send D the new key $K_{B',D}$ that is encrypted using $aK_{AP,D}$, which B' obtains from $Cert_{CA}(D)$ by decrypting it using the AP group key gK_{AP}. D can obtain the new key $K_{B',D}$.

However, before D can check B''s identity, it has to wait until A publishes the TESLA key tK_{A_n} during time slot $n + d$. Once D receives that key, D can check B''s certificate and confirm the identity of B'. Subsequently, D may resume sending its data to the application by securely sending its data to B' using $K_{B',D}$.

Switching between different forwarding nodes doesn't require a sensor node handoff, since a sensor node initially doesn't share a secret key with any forwarding node. However, if the the sensor node is sending data in *assured mode* while it is switches forwarding nodes, the assured service can no longer be provided. In this case the new forwarding node sends an error message to the sensor node. If the sensor node wants to continue sending in *assured service*, it has to reinitiate the service with the new forwarding node.

10.8   Data Origin Authentication

The framework's data origin authentication service enables the application A to check whether the received data are sent by a valid sensor node. The sensor node D uses the secret key that it shares with the application to create a MAC that it appends to the data packet. Depending on the sensitivity of the data or the overall trust in the network, D can decide to use the *Weak Mode* or the more sophisticated *Assured Mode* to send the data. The application authenticates unicast data by appending a MAC created with the secret key it shares with the relevant node. It gains access to the wireless part of the network by appending another MAC that it creates using the key it shares with the relevant access point. Access points forward only authenticated application data to avoid Internet rendered attacks against the wireless devices. Broadcast application data is authenticated using the TESLA protocol.

10.8.1   Sending Sensor Data in Weak Mode

Since there are frequent changes in the network topology, the sensor node D doesn't know if its data will arrive directly at the gateway access point

B or will first be received by a forwarding node C, that forwards it either to another forwarding node C' or to B. Therefore, the format of the data packet must depend only on the three entities that will always be involved in the data sending process (unless a handoff happens): the sensor node D itself, the gateway access point B and the application A. As shown in Algorithm 17, the packet contains three fields in addition to the actual data: two MAC-fields and one encrypted random-number-field. If the forwarding node C receives the packet, C forwards it without doing any modification to the packet. Once the gateway access point B gets it, B checks the first MAC. If it is valid, B removes the MAC and the random-number from the packet, adds a new MAC using the secret key it shares with the application and sends it on. After that, it decrypts the random number, adds one, encrypts the result and sends it back to D. Receiving the result, D can be sure that the data is on the right way to the application. However, this *challenge-response-mechanism* does not enable D to figure out who delivered the packet to the gateway access point. There also is no certainty that a misbehaving forwarding node copied the packet without being detected. A sensor node that wants all nodes on the path to the application to be authentcated has to choose the *assured mode*.

Upon the receipt of the data, the application A checks both MACs and, if both are valid, processes the data. If one of them is invalid, A returns a *data-reject*-message (*dRej*) that includes information about which MAC caused trouble. It appends two MACs that enable B and D to verify the application as the sender of the reject-message.

10.8.2 Sending Sensor Data in Assured Mode

The assured mode provides authenticity along the path of the packet at the cost of additional message exchange, higher computational overhead and less flexibility. A sensor node D that wants to send in assured mode first sends an *assured-data-request* (*asdReq*) that contains an encrypted secret key that will be used to install a shared secret between D and the forwarding nodes along the path. Algorithm 18 shows a case, in which a forwarding node C relays the packets from D to B. Once C receives the request, it will sign the packet and append its *cert* before forwarding it. The gateway access point B that gets the packet first checks the certificate, then the signature and, if both are valid, replies with an *assured-data-confirmation* (*asdConf*) that includes the secret key $K_{C,D}$ encrypted with the forwarding node's public key. The forwarding node extracts $K_{C,D}$ from the confirmation packet and uses it to create and append a MAC to the confirmation message before sending it on to the sensor node. Once the packet reaches the sensor node, it will check the MAC. If the MAC is valid, D can be sure, that its gateway access point trusts the forwarding node.

To make sure that the data takes the authenticated path, the sensor node additionally encrypts the *challenge-response random number* with $K_{C,D}$. In the case that more than one forwarding node lies on the path between the sensor node and its gateway access point, each of them appends its signature to the request before forwarding it in the direction of the access point. The access point will establish the additional needed shared secrets between sensor and forwarding nodes. D will successively encrypt the random number using each of the keys in the appropriate sequence before sending data.

10.9   Evaluation

In [173], L. Zhou and Z. Haas provide a general overview of security challenges and threats in ad hoc networks. In this section we provide basic security and performance analysis for the proposed framework on the basis of their security criteria. Throughout the evaluation we assume the TESLA protocol to be secure and that loose time synchronization exists in the network.

10.9.1   Security Analysis

The use of wireless links renders an ad hoc network susceptible link attacks. Secondly, because mobile nodes may be compromised, malicious attacks have to be considered from outside and inside the network.

Wireless Link Attacks

Since this chapter addresses authentication in hierarchical ad hoc networks, neither application nor sensor data is protected against eavesdropping attacks on the wireless links. However, installing a confidentiality service on top of an existing authentication service is a relatively easy task and is one of the next issues that we will address. The use of message authentication codes in our framework protects all data against malicious modifications and information forgery. In our framework, we address the threat of authorization violation by providing an access control service at the access point level. While we can't prevent intruders from coming into the network and sending packets, we can make it uninteresting for them to do so. The most likely reason for an intruder to use the network's resources is to connect to the Internet, which is prevented by access control at the access point. This access control also restricts battery-consumption attacks by making it impossible to launch such attacks from outside the wireless part of the network. The deletion of packets in the wired part of the network is a threat that we haven't addressed yet.

Compromised Nodes

Since all initial authentication is done by the application that drives the network, compromised sensor nodes can't inflict any damage to the network other than feeding the application with wrong data. Because of the constrained battery resources, denial of service (DoS) attacks launched by compromised sensor nodes are unlikely to happen. Even in the case of such an attack, the application can easily find the origin of the DoS-packets and end the sensor node's trust relationships in the entire network. Compromised forwarding nodes don't have any means to threaten the authentication framework because they don't share a secret with the application. Since the sensor data packets include a challenge-response mechanism, false forwarding of packets or their deletion by a forwarding node will be detected by the sensor node. The framework doesn't provide the possibility to avoid or detect the malicious duplication and distribution of sensor node packets. However, any modification to the data will be detected by the access point or the application and in case the data was sent in *assured mode* the malicious forwarding node can be identified. If a compromised forwarding node stops forwarding data to or from the sensor node or continuously modifies sensor node data that then gets rejected at the access point, the sensor node must find a different network node to connect to (i.e. another forwarding node or its gateway access point). A compromised access point threatens the network's access control mechanism. It can stop forwarding packets in both directions. While the forwarding nodes won't detect the problem, the sensor nodes will notice the lack of certificate renewals and application TESLA keys. Since a sensor node is not able to distinguish between a compromised forwarding node and a compromised access point circumstance, it acts exactly the same in this case. However, for an attacker it is much more complicated to compromise an installed access point than a mobile node. Access points should feature a high degree of physical protection. Further, once an attacker manages to compromise the application, the authentication framework fails. However, if the application itself is compromised there is no use in protecting the sensor devices or data.

10.9.2 Performance Analysis

We now provide basic performance analysis for our authentication framework according to the major performance criteria of [173]. First, since ad hoc networks feature a frequently changing topology, security solutions must be highly adaptable, and secondly, because an ad hoc network may consist of thousands of sensor nodes, the security mechanisms should be scalable.

Adaptability

Our proposed architecture is capable of adapting to meet the authentication needs of the sensors resulting from topology changes. The handoff procedure described in Section 10.7 facilitates the establishment of new trust relationships as nodes move without requiring the participation of the application. This is desirable since it does not burden the application, and hence the application does not serve as a bottleneck. Further, our authentication framework does not require the explicit participation of the forwarding nodes in the authentication of data. We therefore do not have to update any authentication parameters due to the mobility of the forwarding nodes.

Scalability

Since our scenario is application-driven, the amount of resources required by sensors to store their authentication keys remains the same as the number of sensors increases. Although the amount of keys that must be maintained by access points and the application will increase, these entities have more resources to devote to security services.

We have conducted an initial evaluation the amount of time required to a 4096-bit message authentication using SHA-1, and 2048-bit RSA signing using the libtomcrypt library [174]. Using gprof on a Pentium-4 2GHz Linux machine, we measured that SHA-1 required an average of 46 milliseconds to perform, while RSA signing required an average of 2.26 seconds to perform. Although this platform is not the same as a typical sensor node, the timing measurements do allow us to estimate that the RSA operation requires roughly 4900 times more power than performing SHA-1. We are currently conducting a more thorough estimation of power consumption on Cerfcubes [175], which is a typical sensor node device.

In our framework, we have sought to distribute the computational load according to each layer's capabilities. Examination of our protocols reveals that we have not burdened sensor nodes with public key operations. Instead, due to their use of TESLA certificates, the sensor nodes use computationally efficient MACs to perform authentication. However, we have placed more computational burden upon higher-powered access points by requiring them to perform public key cryptographic operations.

10.10   Conclusion

In this chapter, we presented an authentication framework for an application-driven hierarchical ad hoc network. Our framework authenticates incoming nodes, maintains trust relationships during topology changes through a flexible handoff scheme, and provides data origin authentication for sensor data. Further, the presented framework treats nodes according to their resource limitations. In particular, weak sensor nodes are not involved with the creation or validation of public key signatures. Instead, sensor nodes perform runtime entity authentication by the means of TESLA certificates, an alternative to the widely used PKI certificates.

Algorithm:*Sending Sensor Data in* **weak mode**

Result: *The application A can be sure that the received data comes from sensor node D*

1 $D \to C : (data, \{rn\}_{K_{B,D}}, MAC_{K_{B,D}}(data),$
$\qquad MAC_{K_{A,D}}(data))$

2 $C \to B : (data, \{rn\}_{K_{B,D}}, MAC_{K_{B,D}}(data),$
$\qquad MAC_{K_{A,D}}(data))$

3 **if** *($MAC_{K_{B,D}}(data)$ valid)* **then**
$\quad$ $B \to A : (data, MAC_{K_{A,D}}(data),$
$\qquad\qquad MAC_{K_{A,B}}(data))$
$\quad$ $B \to C : (\{rn+1\}_{K_{B,D}})$
$\quad$ $C \to D : (\{rn+1\}_{K_{B,D}})$
end

4 **if** *($MAC_{K_{A,B}}(data)$ valid)* **then**
$\quad$ **if** *($MAC_{K_{A,D}}(data)$ valid)* **then**
$\quad\quad$ A processes data
$\quad$ **else**
$\quad\quad$ $A \to B : (dRej, D, MAC_{K_{A,B}}(dRej, D),$
$\qquad\qquad MAC_{K_{A,D}}(dRej, D))$
$\quad\quad$ $B \to C : (dRej, D, MAC_{K_{B,D}}(dRej, D),$
$\qquad\qquad MAC_{K_{A,D}}(dRej, D))$
$\quad\quad$ $C \to D : (dRej, D, MAC_{K_{B,D}}(dRej, D),$
$\qquad\qquad MAC_{K_{A,D}}(dRej, D))$
$\quad$ **end**
else
$\quad$ $A \to B : (dRej, B, MAC_{K_{A,B}}(dRej, B),$
$\qquad\qquad MAC_{K_{A,D}}(dRej, B))$
$\quad$ $B \to C : (dRej, B, MAC_{K_{B,D}}(dRej, B),$
$\qquad\qquad MAC_{K_{A,D}}(dRej, B))$
$\quad$ $C \to D : (dRej, B, MAC_{K_{B,D}}(dRej, B),$
$\qquad\qquad MAC_{K_{A,D}}(dRej, B))$
end

Algorithm 17: *Sending Sensor Data in* **weak mode**

Algorithm:*Sending Sensor Data in* **assured mode**

Result*: The application A can be sure that the received data comes from sensor node D*

1 $D \rightarrow C : (asdReq, \{K_{C,D}\}_{K_{B,D}})$
$\quad C \rightarrow B : (asdReq, \{K_{C,D}\}_{K_{B,D}},$
$\qquad SIGN_{-K_C}(asdReq), Cert_{TTP}(C))$

2 **if** *(SIGN$_{-K_C}$(asdReq) valid)* **then**
$\quad\quad B \rightarrow C : (asdConf, \{K_{CD}\}_{+K_C})$
$\quad\quad C \rightarrow D : (asdConf, MAC_{K_{C,D}}(asdConf))$
end

3 $D \rightarrow C : (data, \{\{rn\}_{K_{B,D}}\}_{K_{C,D}}, MAC_{K_{B,D}}(data),$
$\qquad MAC_{K_{A,D}}(data))$
$\quad C \rightarrow B : (data, \{rn\}_{K_{B,D}}, MAC_{K_{B,D}}(data),$
$\qquad MAC_{K_{A,D}}(data))$

4 **if** *(MAC$_{K_{B,D}}$(data) valid)* **then**
$\quad\quad B \rightarrow A : (data, MAC_{K_{A,D}}(data),$
$\quad\quad\quad MAC_{K_{A,B}}(data))$
$\quad\quad B \rightarrow C : (\{rn+1\}_{K_{B,D}})$
$\quad\quad C \rightarrow D : (\{rn+1\}_{K_{B,D}})$
end

5 *... similar to weak mode.*

Algorithm 18*: Sending Sensor Data in assured mode*

Abbreviation	Explanation
ID_D	Identification Number of node D
TS_B	Time Stamp issued by node B
$K_{B,D}$	Symmetric Key shared by the nodes B and D
$+K_A$	RSA Public Key of application A
$-K_A$	RSA Private Key of application A
gK_{AP}	Symmetric Key shared by all access points and the application (group key)
tK_{A_n}	TESLA Key of application A, valid in timeslot n
$SIGN_{-K_B}(offer)$	Signature over $offer$ by node B using its private key $-K_B$
$SIGN_{-K_{TTP}}(...)$	Signature over the complete packet by the TTP using $-K_{TTP}$
$MAC_{K_{A,D}}(data)$	Message Authentication Code of $data$ using the key $K_{A,D}$
$MAC_{K_{B,D}}(...)$	Message Authentication Code of the complete packet using the key $K_{B,D}$
iK_D	Initial Key of node D issued by the TTP
aK_{B_n}	Access Point B's Authentication Key disclosed in timeslot n
$aK_{AP,D}$	Sensor Node D's Authentication key needed for handoff with access point
$iCert_{TTP}(D)$	Initial Certificate of node D issued by the TTP
$Cert_{A_n}(B)$	TESLA Runtime Certificate of access point B issued by application A, valid in timeslot n
$Cert_A(D)$	Runtime Certificate of sensor node D issued by application A

TABLE 10.1. Notation used in the protocols in Chapter 10

References

[1] S. Paul, *Multicasting on the Internet and its Applications*, Kluwer Academic, 1998.

[2] S. Keshav, *An Engineering Approach to Computer Networking: ATM Networks, the Internet, and the Telephone Network*, Addison Wesley, 1997.

[3] M. Just, E. Kranakis, D. Krizanc, and P. vanOorschot, "On key distribution via true broadcasting," in *Proc. 2nd ACM Conf. on Computer and Communications Security*, 1994, pp. 81–88.

[4] C. Blundo, L.A. Frota Mattos, and D. R. Stinson, "Multiple key distribution maintaining user anonymity via broadcast channels," *J. Computer Security*, vol. 3, pp. 309–323, 1994.

[5] H. Harney and C. Muckenhirn, "Gkmp specification," Internet Request for Comments 2094, July 1997.

[6] A. Fiat and M. Naor, "Broadcast encryption," *Advances in Cryptology: Crypto '93*, pp. 480–491, 1993.

[7] D.M. Wallner, E.J. Harder, and R.C. Agee, "Key management for multicast: issues and architectures," Internet Draft Report, Sept. 1998, Filename: draft-wallner-key-arch-01.txt.

[8] C. Wong, M. Gouda, and S. Lam, "Secure group communications using key graphs," *IEEE/ACM Trans. on Networking*, vol. 8, pp. 16–30, Feb. 2000.

[9] D. Balenson, D. McGrew, and A. Sherman, "Key management for large dynamic groups: one-way function trees and amortized initialization," Internet Draft Report.

[10] R. Canetti, Juan Garay, Gene Itkis, Daniele Miccianancio, Moni Naor, and Benny Pinkas, "Multicast security: a taxonomy and some efficient constructions," in *IEEE INFOCOM'99*, 1999, pp. 708 –716.

[11] R. Canetti, T. Malkin, and K. Nissim, "Efficient communication-storage tradeoffs for multicast encryption," *Eurocrypt*, pp. 456–470, 1999.

[12] R. Poovendran and J.S. Baras, "An information theoretic approach for design and analysis of rooted tree-based multicast key management schemes," *Advances in Cryptology: Crypto '99*, pp. 624–638, 1999.

[13] A. Menezes, P. vanOorschot, and S. Vanstone, *Handbook of Applied Cryptography*, CRC Press, 1997.

[14] J. Daemen and V. Rijmen, "AES proposal: Rijndael," See http://crsc.nist.gov/encryption/aes, 2000.

[15] ITU-T Recommendation X.509 (1997), "The director: Authentication framework," 1997.

[16] K. Atkinson, *An Introduction to Numerical Analysis*, John Wiley & Sons, 2nd edition, 1989.

[17] G. Golub and C. Van Loan, *Matrix Computations*, The Johns Hopkins University Press, 3rd edition, 1996.

[18] N. Koblitz, *A Course in Number Theory and Cryptography*, Springer-Verlag, 2nd edition, 1994.

[19] W. Diffie and M. Hellman, "New directions in cryptography," *IEEE Trans. on Information Theory*, vol. 22, pp. 644–654, 1976.

[20] I. Ingemarsson, D. Tang, and C. Wong, "A conference key distribution system," *IEEE Transactions on Information Theory*, vol. 28, pp. 714–720, September 1982.

[21] M. Burmester and Y. Desmedt, "A secure and efficient conference key distribution scheme," *Advances in Cryptology- Eurocrypt*, pp. 275–286, 1994.

[22] M. Steiner, G. Tsudik, and M. Waidner, "Diffie-Hellman key distribution extended to group communication," in *Proc. 3rd ACM Conf. on Computer Commun. Security*, 1996, pp. 31–37.

[23] K. Becker and U. Wille, "Communication complexity of group key distribution," in *5th ACM Conf. on Computer Commun. Security*, 1998, pp. 1–6.

[24] G. Ateniese, M. Steiner, and G. Tsudik, "New multiparty authentication services and key agreement protocols," *IEEE Journal on Selected Areas of Communications*, vol. 18, pp. 628 –639, 2000.

[25] M. Steiner, G. Tsudik, and M. Waidner, "Key agreement in dynamic peer groups," *IEEE Transactions on Parallel and Distributed Systems*, vol. 11, pp. 769 –780, 2000.

[26] V. Miller, "Use of elliptic curves in cryptography," *Advances in Cryptology: Crypto '85*, pp. 417–426, 1986.

[27] W. Trappe, Y. Wang, and K.J.R. Liu, "Group key agreement using divide-and-conquer strategies," in *Conference on Information Sciences and Systems, The John's Hopkins University*, March 2001.

[28] W. Trappe, Y. Wang, and K.J.R. Liu, "Establishment of conference keys in heterogenous networks," in *Proceedings of the IEEE Int. Conference on Communications*, 2002, pp. 2201–2205.

[29] A. Oppenheim and R. Schafer, *Discrete-time Signal Processing*, Prentice Hall, 1989.

[30] T. Cover and J. Thomas, *Elements of Information Theory*, John Wiley and Sons, 1991.

[31] D. Huffman, "A method for the construction of minimum redundancy codes," *Proc. of IRE*, vol. 40, pp. 1098–1101, 1952.

[32] D. A. Lelewer and D. S. Hirschberg, "Data compression," *ACM Computing Surveys*, vol. 19, pp. 261–296, 1987.

[33] D. E. Knuth, *The Art of Computer Programming, Vol. 3, Sorting and Searching*, Addison Wesley, 1973.

[34] T. Cormen, C. Leiserson, and R. Rivest, *Introduction to Algorithms*, Mc. Graw Hill, 1998.

[35] A. Turping and A. Moffat, "Practical length-limited coding for large alphabets," *Computer Journal*, vol. 38, pp. 339–347, 1995.

[36] D. C. Van Voorhis, "Constructing codes with bounded codeword lengths," *IEEE Transactions on Information Theory*, vol. 20, pp. 288–290, March 1974.

[37] L. Larmore and D. Hirschberg, "A fast algorithm for optimal length-limited Huffman codes," *Journal of the ACM*, vol. 37, pp. 464–473, July 1990.

[38] R. Milidiu and E. Laber, "The warm-up algorithm: a Lagrangian construction of length restricted Huffman codes," *SIAM Journal of Computation*, vol. 30, pp. 1405–1426, 2000.

[39] M. R. Garey, "Optimal binary search trees with restricted maximal depth," *SIAM Journal of Computing*, vol. 3, pp. 101–110, June 1974.

[40] E. Gilbert, "Codes based on inaccurate source probabilities," *IEEE Trans. on Inform. Theory*, vol. 17, pp. 304–314, 1971.

[41] H. Murakami, S. Matsumoto, and H. Yamamoto, "Algorithm for construction of variable length code with limited maximum word length," *IEEE Transactions on Communications*, vol. 32, pp. 1157–1159, Oct. 1984.

[42] B. Fox, "Discrete optimization via marginal analysis," *Management Science*, vol. 13, pp. 210–216, 1966.

[43] L. A. Wolsey, *Integer Programming*, John Wiley and Sons, 1998.

[44] T. C. Hu, *Integer Programming and Network Flows*, Addison Wesley, 1969.

[45] G. L. Nemhauser and L. A. Wolsey, *Integer and Combinatorial Optimization*, John Wiley and Sons, 1999.

[46] A. H. Land and A. G. Doig, "An automatic method for solving discrete programming problems," *Econometrica*, vol. 28, pp. 497–520, 1960.

[47] E. L. Lawler and D. E. Wood, "Branch-and-bound methods: A survey," *Operations Research*, vol. 14, pp. 699–719, 1966.

[48] D. C. Little, K. G. Murty, D. W. Sweeney, and C. Karel, "An algorithm for the traveling salesman problem," *Operations Research*, vol. 11, pp. 972–989, 1963.

[49] W. Trappe, J. Song, R. Poovendran, and K.J.R. Liu, "Key distribution for secure multimedia multicasts via data embedding," in *IEEE Int. Conference on Acoustics, Speech, and Signal Processing*, 2001, pp. 1449–1452.

[50] T. Nemetz, "On the word-length of Huffman codes," *Probl. Contr. and Inform. Theory*, vol. 9, pp. 231–242, 1980.

[51] F. Fabris, A. Sgarro, and R. Pauletti, "Tunstall adaptive coding and miscoding," *IEEE Trans. on Inform. Theory*, vol. 42, pp. 2167–2180, 1996.

[52] I. Ingemarsson, D. Tang, and C. Wong, "A conference key distribution system," *IEEE Transactions on Information Theory*, vol. 28, pp. 714–720, Sep. 1982.

[53] D. G. Steer, L. Strawczynski, W. Diffie, and M. Wiener, "A secure audio teleconference system," in *Proceedings on Advances in cryptology*. 1990, pp. 520–528, Springer-Verlag New York, Inc.

[54] M. Burmester and Y. Desmedt, "A secure and efficient conference key distribution scheme," *Advances in Cryptology- Eurocrypt*, pp. 275–286, 1994.

[55] W. Diffie and M. Hellman, "New directions in cryptography," *IEEE Trans. on Information Theory*, vol. 22, pp. 644–654, 1976.

[56] M. Steiner, G. Tsudik, and M. Waidner, "Diffie-Hellman key distribution extended to group communication," in *Proceedings of the 3rd ACM conference on Computer and communications security*. 1996, pp. 31–37, ACM Press.

[57] G. Ateniese, M. Steiner, and G. Tsudik, "Authenticated group key agreement and friends," in *ACM Conference on Computer and Communication Security*, 1998, pp. 17–26.

[58] M. Steiner, G. Tsudik, , and M. Waidner, "Key agreement in dynamic peer groups," *IEEE TRANSACTIONS ON PARALLEL AND DISTRIBUTED SYSTEMS*, vol. 11, no. 8, pp. 769–780, Aug 2000.

[59] Y. Kim, A. Perrig, and G. Tsudik, "Tree-based group key agreement," *ACM Transactions on Information and System Security*, vol. 7, no. 1, pp. 60–96, Feb. 2004.

[60] Y. Mao, Y. Sun, M. Wu, and K. J. R. Liu, "Dynamic join-exit amortization and scheduling for time-efficient group key agreement," in *IEEE INFOCOM*, 2004.

[61] W. Trappe, Y. Wang, and K.J.R. Liu, "Establishment of conference keys in heterogeneous networks," in *proceedings of IEEE International Conference on Communications*, 2002, vol. 4, pp. 2201–2205.

[62] Jack Snoeyink, Subhash Suri, and George Varghese, "A lower bound for multicast key distribution," in *IEEE INFOCOM*, 2001.

[63] Y. Mao, Y. Sun, M. Wu, and K. J. Ray Liu, "Join-exit scheduling for contributory group key agreement," *IEEE/ACM Transactions on Networking*, vol. 14, pp. 1128–1140, Oct 2006.

[64] W. Yu, Y. Sun, and K. J. Ray Liu, "Minimization of rekeying cost for contributory group communications," in *Proc. IEEE Globecom'05*, Nov-Dec 2005.

[65] W. Yu, Y. Sun, and K. J. Ray Liu, "Optimizing rekeying cost for contributory group key agreement schemes," *accepted, IEEE Trans. on Dependable and Secure Computing*, 2007.

[66] G. Tsudik Y. Kim, A. Perrig, "Simple and fault-tolerant key agreement for dynamic collaborative groups," in *Proceedings of the 7th ACM conference on Computer and communications security*, 2000, pp. 235–244.

[67] L.R. Dondeti, S. Mukherjee, and A. Samal, "DISEC: a distributed framework for scalable secure many-to-many communication," in *Proceedings of Fifth IEEE Symposium on Computers and Communications*, 2000, pp. 693–698.

[68] K. Becker and U. Wille, "Communication complexity of group key distribution," in *Proceedings of 5th ACM Conf. on Computer Commun. Security*, 1998, pp. 1–6.

[69] J. L. Hennessy and D. A. Patterson, *Computer Architecture: a Quantitative Approach*, Morgan Kaufmann publishers, second edition, 1996.

[70] T. H. Corman, C. E. Leiserson, and R. L. Rivest, *Introduction to Algorithms*, The MIT Press and McGraw-Hill Book Company, second edition, 2001.

[71] "Mbone user activity data," ftp://ftp.cc.gatech.edu/people/kevin/release-data, 2003.

[72] K. Almeroth and M. Ammar, "Multicast group behavior in the internet's multicast backbone (MBone)," *IEEE Communications*, vol. 35, pp. 224–229, June 1997.

[73] K Sara and K. C. Almeroth, "Supporting multicast deployment efforts: a survey of tools for multicast monitoring," *Journal of High Speed Networks*, vol. 9, no. 3-4, pp. 191–211, 2000.

[74] M. Bellare and P. Rogaway, "Random oracles are practical: A paradigm for designing efficient protocols," in *ACM Conference on Computer and Communication Security*, 1993.

[75] C. Diot, B.N. Levine, B. Lyles, H. Kassem, and D. Balensiefen, "Deployment issues for the IP multicast service and architecture," *IEEE Network*, vol. 14, pp. 78–88, Jan.-Feb 2000.

[76] A. Acharya and B.R. Badrinath, "A framework for delivering multicast messages in networks with mobile hosts," *Journal of Special Topics in Mobile Networks and Applications*, vol. 1, no. 2, pp. 199–219, Oct. 1996.

[77] H-S Shin and Y-J Suh, "Multicast routing protocol in mobile networks," *Proc. IEEE International Conference on Communications*, vol. 3, pp. 1416 –1420, June 2000.

[78] M.J. Moyer, J.R. Rao, and P. Rohatgi, "A survey of security issues in multicast communications," *IEEE Network*, vol. 13, no. 6, pp. 12–23, Nov.-Dec. 1999.

[79] W. Trappe, J. Song, R. Poovendran, and K.J.R. Liu, "Key distribution for secure multimedia multicasts via data embedding," *Proc. IEEE ICASSP'01*, pp. 1449–1452, May 2001.

[80] M. Waldvogel, G. Caronni, D. Sun, N. Weiler, and B. Plattner, "The VersaKey framework: Versatile group key management," *IEEE Journal on selected areas in communications*, vol. 17, no. 9, pp. 1614–1631, Sep. 1999.

[81] Y. Sun, W. Trappe, and K. J. R. Liu, "A scalable multicast key management scheme for heterogeneous wireless networks," *IEEE/ACM Trans. on Networking*, vol. 12, no. 4, pp. 653–666, Aug. 2004.

[82] K. Brown and S. Singh, "RelM: Reliable multicast for mobile networks," *Computer Communication*, vol. 2.1, no. 16, pp. 1379–1400, June 1996.

[83] E. Ha, Y. Choi, and C. Kim, "A multicast-based handoff for seamless connection in picocellular networks," *Proc. IEEE Asia Pacific Conference on Circuits and Systems*, pp. 167–170, Nov. 1996.

[84] Universal Mobile Telecommunications System (UMTS) Technical Specification, Digital cellular telecommunications system (Phase 2+ (GSM)), "Network architecture," 3GPP TS 23.002 version 5.9.0 Release 5, 2002-12.

[85] J.E. Wieselthier, G.D. Nguyen, and A. Ephremides, "On the construction of energy-efficient broadcast and multicast trees in wireless networks," *Proc. IEEE INFOCOM'00*, vol. 2, pp. 585–594, March 2000.

[86] L. Gong and N. Shacham, "Multicast security and its extension to a mobile environment," *Wireless Networks*, vol. 1, no. 3, pp. 281–295, 1995.

[87] M. Hauge and O. Kure, "Multicast in 3G networks: employment of existing IP multicast protocols in umts," in *Proceedings of the 5th ACM international workshop on Wireless mobile multimedia*. 2002, pp. 96–103, ACM Press.

[88] S. Paul, *Multicast on the Internet and its applications*, Kluwer Academic Publishers, 1998.

[89] "Mlisten," available at www.cc.gatech.edu/computing/Telecomm. mbone.

[90] K. Almeroth and M. Ammar, "Collecting and modeling the join/leave behavior of multicast group members in the mbone," in *Proc. High Performance Distributed Computing (HPDC'96); Syracuse, New York*, 1996, pp. 209–216.

[91] G.K. Zipf, *Human Behavior and the Principle of Least Effort*, Addison-Wesley Press, 1949.

[92] A. Leon-Garcia, *Probability and Random Processes for Electrical Engineering*, Addison Wesley, 2nd edition, 1994.

[93] M. M. Zonoozi and P. Dassanayake, "User mobility modeling and characterization of mobility patterns," *IEEE Journal on Selected Areas in Communications*, vol. 15, no. 7, pp. 1239–1252, Sep. 1997.

[94] A. Puri and T. Chen, *Multimedia Systems, Standards, and Networks*, Marcel Dekker Inc., 2000.

[95] U. Horn, K. Stuhlmller, M. Link, and B. Girod, "Robust internet video transmission based on scalable coding and unequal error protection," *Image Communication*, vol. 15, pp. 77–94, Sept 1999.

[96] H. Zheng and K.J.R. Liu, "Optimization approaches for delivering multimedia services over digital subscriber lines," *IEEE Signal Processing Magazine*, vol. 17, pp. 44–60, July 2000.

[97] W. Trappe and L.C. Washington, *Introduction to Cryptography with Coding Theory*, Prentice Hall, 2002.

[98] C. Herpel, A. Eleftheriadis, and G. Franceschini, "MPEG-4 systems: elementary stream management and delivery," in *Multimedia Systems, Standards, and Networks*, A. Puri and T. Chen, Eds., pp. 367–405. Marcel Dekker Inc., 2000.

[99] S. Floyd, V. Jacobson, C. Liu, S. McCanne, and L. Zhang, "A reliable multicast framework for light-weight sessions and application level framing," *IEEE/ACM Transactions on Networking*, vol. 5, pp. 784–803, 1997.

[100] J. Lin and S. Paul, "RMTP: A reliable multicast transport protocol," in *INFOCOM*, San Francisco, CA, Mar 1996, pp. 1414–1424.

[101] S. Paul, K. K. Sabnani, J. Lin, and S. Bhattacharyya, "Reliable multicast transport protocol (RMTP)," *IEEE Journal of Selected Areas in Communications*, vol. 15, pp. 407–421, 1997.

[102] R. Poovendran and J.S. Baras, "An information-theoretic approach for design and analysis of rooted-tree-based multicast key management schemes," *IEEE Trans. on Information Theory*, vol. 47, pp. 2824 –2834, 2001.

[103] J. Song, R. Poovendran, W. Trappe, and K.J.R. Liu, "A dynamic key distribution scheme using data embedding for secure multimedia multicast," in *Proceedings of SPIE 2001 Security and Watermarking for Multimedia*, San Jose, CA, 2001.

[104] I. Cox, J. Kilian, F. Leighton, and T. Shamoon, "Secure spread spectrum watermarking for multimedia," *IEEE Tran. on Image Proc.*, vol. 6(12), pp. 1673–1687, December 1997.

[105] F. Hartung and B. Girod, "Digital watermarking of MPEG-2 coded video in the bitstream domain," *IEEE Int. Conf. Accostic Speech and Signal Proc. '97*, pp. 2621–2624, 1997.

[106] C. Podilchuk and W. Zeng, "Image adaptive watermarking using visual models," *IEEE Journal on Selected Areas in Communications*, vol. 16(4), pp. 525–540, May 1998.

[107] A. Westfeld and G. Wolf, "Steganography in a video conferencing system," in *Proc. 2nd International Workshop on Information Hiding*, 1998, pp. 32–47.

[108] J. Song and K. J. R. Liu, "A data embedding scheme for H.263 compatible video coding," *IEEE ISCAS*, vol. 4, pp. 390–393, June 1999.

[109] J. Song and K. J. R. Liu, "A data embedded video coding scheme for error-prone channels," *IEEE Trans. on Multimedia*, vol. 3, pp. 415–423, Dec. 2001.

[110] J. Nonnenmacher, E. W. Biersack, and D. Towsley, "Parity-based loss recovery for reliable multicast transmission," *IEEE/ACM Trans. Networking*, vol. 5, pp. 349–361, Aug 1998.

[111] D. Stinson, *Cryptography: Theory and Practice*, CRC Press, 1995.

[112] J. B. Conway, *Functions of One Complex Variable*, Springer-Verlag, 2nd edition, 1978.

[113] ITU-T Rec. H263, "Version 2, video coding for low bitrate communication," Jan. 1998.

[114] A. Puri and T. Chen, *Multimedia Systems, Standards, and Networks*, Marcel Dekker Inc, 2000.

[115] Y. Sun and K. J. Ray Liu, "Dynamic key graph for hierarchical access control in secure group communications," *IEEE/ACM Transactions on Networking*, Feb 2008.

[116] Y. Sun and K. J. Ray Liu, "Scalable hierarchical access control in secure group communications," in *Proc. IEEE INFOCOM'04*, March 2004.

[117] D. McGrew and A. Sherman, "Key establishment in large dynamic groups using one-way function trees," Technical Report 0755, TIS Labs at Network Associates, Inc., Glenwood, MD, May 1998.

[118] A. Perrig, D. Song, and D. Tygar, "ELK, a new protocol for efficient large-group key distribution," in *Proc. IEEE Symposium on Security and Privacy*, 2001, pp. 247 –262.

[119] Y. R. Yang, X. S. Li, X. B. Zhang, and S. S. Lam, "Reliable group rekeying: a performance analysis," *Proc. of the 2001 conference on applications, technologies, architectures, and protocols for computer communications*, pp. 27 – 38, August 2001.

[120] B. Sun, W. Trappe, Y. Sun, and K.J.R. Liu, "A time-efficient contributory key agreement scheme for secure group communications," *Proc. of IEEE International Conference on Communication*, vol. 2, pp. 1159 –1163, 2002.

[121] Mikhail J. Atallah, Keith B. Frikken, and Marina Blanton, "Dynamic and efficient key management for access hierarchies," in *CCS '05: Proceedings of the 12th ACM conference on Computer and communications security*, New York, NY, USA, 2005, pp. 190–202.

[122] Q. Zhang and Y. Wang, "A centralized key management scheme for hierarchical access control," in *Proceedings on IEEE Global Telecommunications Conference (Globecom'04)*, 2004.

[123] Y. Sun and K. J. Ray Liu, "Securing dynamic membership information in multicast communications," in *Proc. IEEE INFOCOM'04*, March 2004.

[124] Y. Sun and K. J. Ray Liu, "Analysis and protection of dynamic membership information for group key distribution schemes," *IEEE Journal on Information Forensics and Security*, June 2007.

[125] K. Almeroth and B. Quinn, "Ip multicast applications: Challenges and solutions," IETF Draft, November 1998, Filename: draft-quinn-multicast-apps-00.txt.

[126] W. Trappe, J. Song, R. Poovendran, and K.J.R. Liu, "Key management and distribution for secure multimedia multicast," *IEEE Trans. on Multimedia*, vol. 5, pp. 544–557, 2003.

[127] "http://ftp.cc.gatech.edu/people/kevin/release-data,".

[128] Y. Amir, G. Ateniese, D. Hasse, Y. Kim, C. Nita-Rotaru, T. Schlossnagle, J. Schultz, J. Stanton, and G. Tsudik, "Secure group communication in asynchronous networks with failures: Integration and experiments," in *Proceedings of IEEE ICDCS 2000*, April 2000.

[129] G. H. Chiou and W. T. Chen, "Secure broadcasting using the secure lock," *IEEE Trans. Software Eng.*, vol. 15, pp. 929–934, Aug 1989.

[130] S. Mittra, "Iolus: A framework for scalable secure multicasting," in *Proc. ACM SIGCOMM '97*, 1997, pp. 277–288.

[131] S. Banerjee and B. Bhattacharjee, "Scalable secure group communication over IP multicast," *JSAC Special Issue on Network Support for Group Communication*, vol. 20, no. 8, pp. 1511 –1527, Oct. 2002.

[132] Y. Sun, W. Trappe, and K.J.R. Liu, "An efficient key management scheme for secure wireless multicast," *Proc. of IEEE International Conference on Communication*, vol. 2, pp. 1236–1240, 2002.

[133] S. Setia, S. Koussih, S. Jajodia, and E. Harder, "Kronos: A Scalable Group Re-keying Approach for Secure Multicast," in *2000 IEEE Symposium on Security and Privacy*. IEEE, May 2000, pp. 215–218, Oakland, CA.

[134] M. Reed, P. Syverson, and D. Goldschlag, "Anonymous connections and onion routing," *IEEE journal on selected areas in communications*, vol. 16, pp. 482–494, May 1998.

[135] M. Steiner, G. Tsudik, and M. Waidner, "CLIQUES: a new approach to group key agreement," in *Proceedings of the 18th International Conference on Distributed Computing Systems*, May 1998, pp. 380–387.

[136] R.E. Newman-Wolfe and B.R. Venkatraman, "High level prevention of traffic analysis," in *Proceedings of Seventh Annual Computer Security Applications Conference*, Dec. 1991, pp. 102–109.

[137] R. Gennaro and P. Rohatgi, "How to sign digital streams," *Advances in Cryptology: Crypto '97*, 1997.

[138] C. K. Wong and S. Lam, "Digital signatures for flows and multicasts," *IEEE/ACM Trans. On Networking*, pp. 502–513, 1999.

[139] P. Rohatgi, "A compact and fast hybrid signature scheme for multicast packet authentication," in *6th ACM Conference on Computer and Communications Security*, 1999, pp. 93–100.

[140] J.M. Park, E.K.P. Chong, and H.J. Siegel, "Efficient multicast packet authentication using signature amortization," in *IEEE Symposium on Security and Privacy*, 2002, pp. 227–240.

[141] J.M. Park, E.K.P. Chong, and H.J. Siegel, "Efficient multicast stream authentication using erasure codes," *ACM Transactions on Information and System Security*, vol. 6, no. 2, pp. 258–285, 2003.

[142] C. Karlof, N. Sastry, Y. Li, A. Perrig, and D. Tygar, "Distillation codes and applications to dos resistant multicast authentication," in *Network and Distributed System Security Symposium*, 2004.

[143] A. Lysyanskaya, R. Tamassia, and N. Triandopoulos, "Multicast authentication in fully adversarial networks," in *IEEE Symposium on Security and Privacy*, 2004, pp. 241–255.

[144] R. Canetti, J. Garay, G. Itkis, D. Micciancio, M. Naor, and B. Pinkas, "Multicast security: A taxonomy and some efficient constructions," in *INFOCOMM'99*, 1999.

[145] A. Perrig, "The BiBa one-time signature and broadcast authentication protocol," in *Eighth ACM Conference on Computer and Communication Security*, November 2001, pp. 28–37.

[146] S. Xu and R. Sandhu, "Authenticated multicast immune to denial-of-service attack," in *ACM Symposium on Applied Computing*, 2002, pp. 196–200.

[147] S. Cheung, "An efficient message authentication scheme for link state routing," in *Proceedings of the 13th Annual Computer Security Applications Conference*, December 1997, pp. 90–98.

[148] R. Anderson, F. Bergadano, B. Crispo, J. Lee, C. Manifavas, and R. Needham, "A new family of authentication protocols," *ACMOSR: ACM Operating Systems Review*, vol. 32, no. 4, pp. 9–20, 1998.

[149] F. Bergadano, D. Cavagnino, and B. Crispo, "Chained stream authentication," in *Proceedings of the 7th Annual Workshop on Selected Areas in Cryptography*, August 2000, pp. 144–157.

[150] F. Bergadano, D. Cavagnino, and B. Crispo, "Individual single source authentication on the mbone," in *IEEE International Conference on Mutlimedia & Expo (ICME)*, August 2000, pp. 541–544.

[151] B. Briscoe, "FLAMeS: Fast, loss-tolerant authentication of multicast streams," *Technical report, BT Research*, 2000.

[152] A. Perrig, R. Canetti, D. Song, J. D. Tygar, and B. Briscoe, "TESLA: Multicast source authentication transform introduction," *IETF working draft, draft-ietf-msec-tesla-intro-01.txt*.

[153] A. Perrig, R. Canetti, D. Song, and J. D. Tygar, "Efficient and secure source authentication for multicast," in *Proceedings of Network and Distributed System Security Symposium*, February 2001.

[154] A. Perrig, R. Canetti, J. D. Tygar, and D. Song, "The TESLA broadcast authentication protocol," *RSA Cryptobytes*, vol. 5, no. 2, pp. 2–13, 2002.

[155] A. Perrig, R. Canetti, J. D. Tygar, and D. Song, "Efficient authentication and signing of multicast streams over lossy channels," in *IEEE Symposium on Security and Privacy*, 2000, pp. 56–73.

[156] C. Kaufman, R. Perlman, and M. Speciner, *Network Security: Private Communication in a Public World*, Prentice Hall, 1995.

[157] F.H.P. Fitzek and M. Reisslein, "MPEG-4 and H.263 video traces for network performance evaluation," *IEEE Network*, vol. 15, no. 6, pp. 40–54, 2001.

[158] D. Gambetta, *Trust: Making and Breaking Cooperative Relations*, Basil Blackwell, Oxford, 1988.

[159] D. Niculescu and B. Nath, "Dv-based positioning in ad hoc networks," *Telecommunication Systems*, pp. 267–280, 2003.

[160] D. Mills, "Network time protocol (version 3) specification, implementation and analysis," *IETF RFC*, http://www.ietf.org/rfc/rfc1305.txt.

[161] A. Gersho and R.M. Gray, *Vertor Quantization and Signal Compression*, Kluwer Academic, 1992.

[162] D. Bertsekas and R. Gallager, *Data Networks*, Prentice Hall, second edition, 1992.

[163] A. Perrig, R. Szewczyk, D. Tygar, V. Wen, and D. Culler, "SPINS: security protocols for sensor networks," *Wireless Networks*, vol. 8, no. 5, pp. 521–534, 2002.

[164] A. Weimerskirch and G. Thonet, "A distributed light-weight authentication model for ad-hoc networks," in *The 4th International Conference on Information Securtiy and Cryptology (ICISC 2001)*, December 2001.

[165] L. Venkatraman and D. Agrawal, "A novel authentiaction scheme for ad hoc networks," in *IEEE Wireless Communications and Networking Conference (WCNC 2000)*, 2000, vol. 3, pp. 1268–1273.

[166] J. Kong, H. Luo, K. Xu, D. Gu, M. Gerla, and S. Lu, "Adaptive security for multi-layer ad-hoc networks," *Special Issue of Wireless Communications and Mobile Computing*, 2002.

[167] P. Gupta and P. Kumar, "The capacity of wireless networks," *IEEE Transactions on Information Theory IT 2000*, vol. IT-46(2), pp. 388–404, 2000.

[168] P. Gupta and P. Kumar, "Internets in the sky: the capacity of three dimensional wireless networks," *Communications in Information Systems*, vol. 1(1), pp. 33–50, 2001.

[169] S. Zhao, K. Tepe, I. Seskar, and D. Raychaudhuri, "Routing protocols for self-organizing hierarchical ad-hoc wireless networks," in *IEEE Sarnoff 2003 Symposium*, 2003.

[170] D. Johnson, D. Maltz, and J. Broch, "DSR: The dynamic source routing protocol for multihop wireless ad hoc networks," in *Ad Hoc Networking, edited by Charles E. Perkins*. 2001, pp. 139–172, Addison-Wesley.

[171] P. R. Zimmermann, *The official PGP user's guide*, MIT Press, 1995.

[172] M. Bellare, R. Canetti, and H. Krawczyk, "Keying hash functions for message authentication," *Advances in Cryptology - Crypto '96*, vol. 1109, pp. 1–15, 1996, Lecture Notes in Computer Science.

[173] L. Zhou and Z. Haas, "Securing ad hoc networks," *IEEE Network*, vol. 13, no. 6, pp. 24–30, 1999.

[174] "Libtomcrypt," www.libtomcrypt.org.

[175] "Intrinsyc product page," www.intrinsyc.com/products/cerfcube.

Index